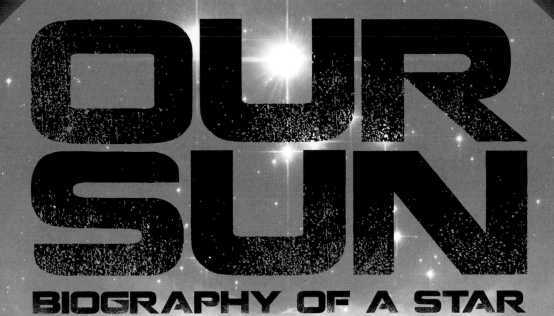

OUR SUN
BIOGRAPHY OF A STAR

CHRISTOPHER COOPER

FOREWORD BY
DAVID SPERGEL, PhD
CHAIR, DEPARTMENT OF ASTROPHYSICAL SCIENCES, PRINCETON UNIVERSITY

PREFACE BY
MADHULIKA GUHATHAKURTA, PhD
LEAD SCIENTIST, LIVING WITH A STAR PROGRAM, NASA

Race Point
PUBLISHING

A division of Book Sales, Inc.
276 Fifth Avenue, Suite 206
New York, New York 10001

RACE POINT PUBLISHING and the distinctive Race Point Publishing logo
are trademarks of Book Sales, Inc.

© 2013 by The Book Shop, Ltd.
7 Peter Cooper Road
New York, NY 10010

This 2013 edition published by Race Point Publishing
by arrangement with The Book Shop, Ltd.

EDITOR, PHOTO RESEARCHER Linda C. Falken
DESIGNER Tim Palin Creative

ISBN-13: 978-1-937994-19-8

Printed in China

2 4 6 8 10 9 7 5 3 1

www.racepointpub.com

CONTENTS

FOREWORD

Why study the Sun? Why explore the planets of our solar system? Why try to understand the properties of the rich variety of stars and planets that fill the Milky Way, our home galaxy? Why peer out in space and back in time and observe the earliest galaxies? Why characterize the leftover heat from the big bang? For me, there are both practical and poetical answers to these questions.

The same laws of physics are valid throughout the universe; thus, by studying the extreme environments of space, we can gain insights into the workings of nature. During the solar eclipse of 1868, the French astronomer Jules Janssen and the British astronomer Norman Lockyer first observed helium in the spectrum of the Sun. Today, helium is used in applications ranging from MRI scanners in hospitals to filling children's balloons. While we do not know the nature of the dark matter that fills our galaxy, perhaps someday our descendants will put it to practical use.

Astronomical research, like all basic research, can also sometimes lead to surprising spin-off technologies. In the early 1990s, John O'Sullivan was searching for radio waves from accreting black holes. He and his team devised a novel computer chip that would let them clear up the radio signals and reduce interference. This chip is an important part of the technology behind WiFi, a technology that is revolutionizing how we interact with our computers, phones, cars, and even our homes.

The space environment can also affect our lives directly and dramatically. A comet once wiped out most of the life on this planet and will likely do so again sometime in the next few hundred million years. Solar flares from the Sun regularly disrupt not only satellite communications but also our electric power grids. The Sun's variations likely have dramatic effects on our climate.

While these practical applications are important, most astronomers do not study the stars so that they can produce new technologies or find a new source of levitation for balloons. We try to understand the universe because of its aesthetic and mathematical beauty. The rich undulating structure of a solar flare or the spidery filaments of a nebula are stunning images. However, the deep mathematical symmetries of the underlying physical principles that govern the formation of spiral arms in a distant galaxy, or the behavior of convective cells that pockmark the Sun, are even more enchanting than the pretty pictures to those of us fortunate enough to study astrophysics. The underlying and universal laws of nature appear to be simultaneously both simple and capable of producing tremendous complexity.

We all know that it is dangerous to stare at the Sun. However, there is a reward associated with the struggle of trying to comprehend its nature. As Edmund Burke, the English philosopher wrote in 1759: "The passion caused by the great and sublime in nature is astonishment, and astonishment is that state of the soul in which all its motions are suspended, with some degree of horror. The mind is so entirely filled with its object that it cannot entertain any other, nor reason on that object which fills it. Astonishment is the effect of the sublime in its highest degree." May your life be touched by moments of astonishment!

David Spergel, PhD
Chair, Department of Astrophysical Sciences, Princeton University

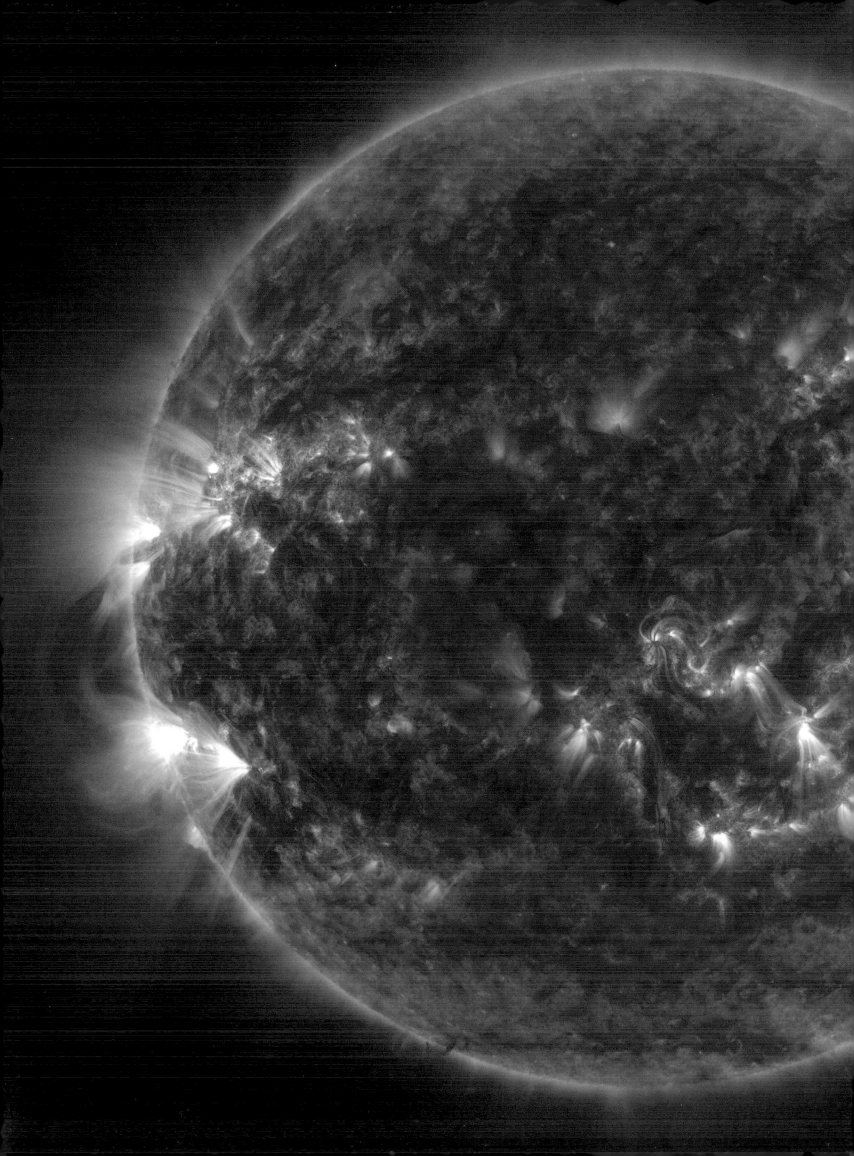

PREFACE

As the lead scientist for NASA's Living With a Star program, I have read hundreds of books about the Sun, ranging from technical tomes for experts to comic books for kids. In short, I've seen it all. Or so I thought until I picked up *Our Sun: Biography of a Star* by Christopher Cooper. His fresh, compelling storytelling made me realize that there is something new under the Sun after all.

Opening the book, the first thing that strikes the reader is the volume's visual appeal. The author recognizes that the Sun is not just a blinding spot of white light in the sky, but rather a profoundly beautiful object. "NASA's Solar Dynamics Observatory [is] producing images of the Sun unlike anything anyone had ever seen before. . . . Words cannot do them justice," writes Cooper. This very correct attitude is reflected in the visual beauty of the book's pictography and layout. Every turn of the page brings another compelling image arranged in storytelling order. The overall effect is to propel the reader forward, flipping pages to see what comes next.

But this book is not just eye candy, it's brain candy. The images and illustrations are selected with the skill of an experienced teacher. You can learn a great deal about the Sun—and the universe it burns in—merely by flipping through the book and looking at the pictures. Textbooks dare you to pick them up; this book dares you to put it down. I could not.

It is really unfair to compare Cooper's biography of the Sun to a textbook, because it is such a different thing. A textbook doles out information in disconnected morsels small enough to memorize. Cooper's tale, on the other hand, is holistic, with multiple themes resonating in every chapter. It is hard to go more than a few pages without experiencing an "ah-ha moment" as some new link is forged across the "void" of space and time or art and science.

Cooper deftly weaves the story of the Sun into the tapestry of human experience, with threads ranging from the trivial to the profound. Humans are creatures of the Sun. We read by its reflected light, we receive its warmth on our skin, we eat food that grew in sunlight. Human biology has evolved to depend on sunlight. In a fascinating aside, the author explores impacts that might be related to humans' declining exposure to their natural habitat—that is, the sunlit outdoors.

As a child growing up in middle-class India, I drove my father crazy, peppering him with constant questions about our place in the cosmos. "Where do we come from?" I would ask. Being more than just a banker—he was also a philosopher and mathematician—he would respond with reason and logic: "Look at a circle. Can you tell me where the end is?" The circular form, in the shape of the Sun, became the gravitational point of my career in heliophysics.

Cooper's story of the Sun begins in the beginning, which he treats much as my father did: "Speaking of 'the beginning' is a bit misleading," Cooper writes. "The prevailing theory is that sometime over 13.79 billion years ago, the universe was an infinitely dense, infinitely hot, and inconceivably tiny speck. . . . No one knows how long the speck had existed—or even if time existed—prior to the Big Bang. The universe might have been there forever, which makes defining 'the beginning' a bit like identifying the start of a circle. Instead, we must choose a point and talk about time relative to that point. For our purposes, the Big Bang is as good a starting point as any."

This is the kind of book my father would have given to his curious daughter—and she would have loved it then, just as she does today.

Madhulika Guha Thakurta

Madhulika Guhathakurta, PhD
Lead Scientist, Living With a Star Program, NASA

The views expressed in this foreword are purely those of Dr. Guhathakurta.

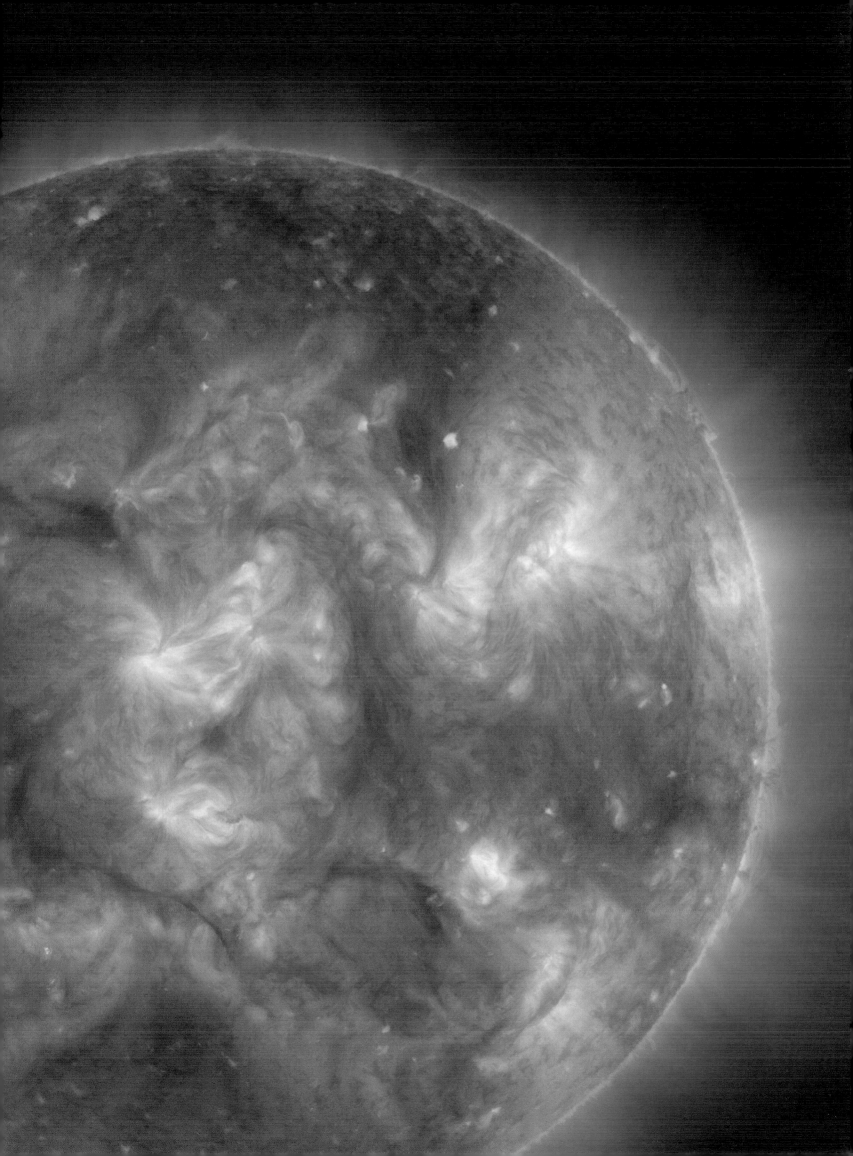

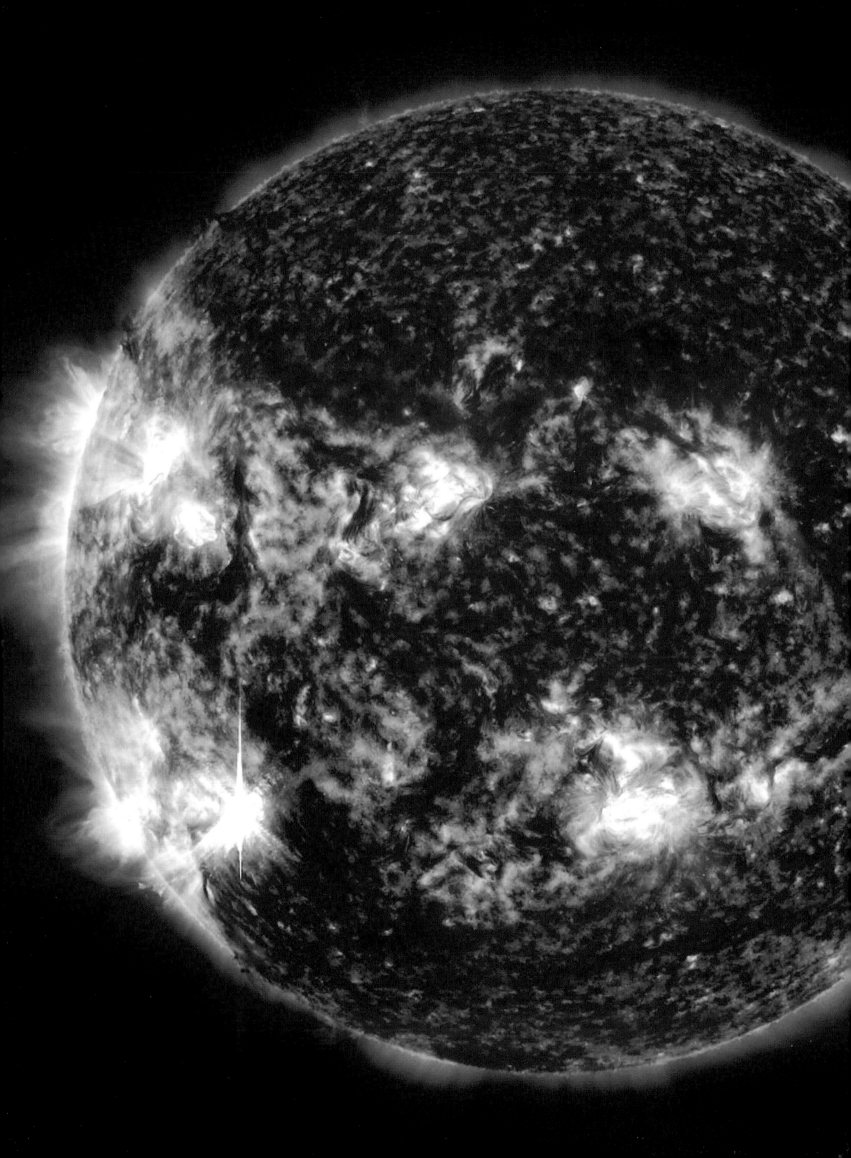

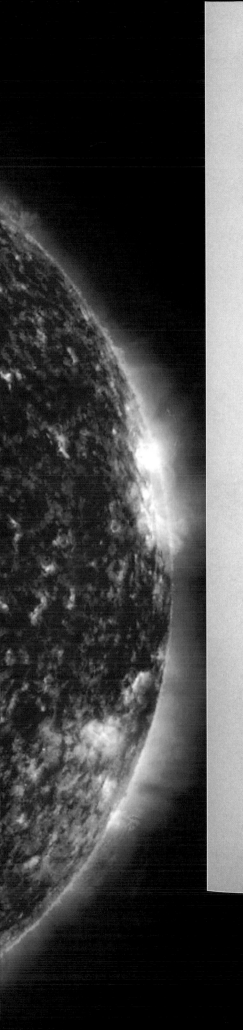

INTRODUCTION

Our Sun is one star among some 300 billion in the galaxy. Our galaxy is one among more than 300 billion in the universe. With a vastness this incomprehensible, it is easy to feel like we are mere specks of sand on an endless shore. But our Sun is special. Though roughly 150 million kilometers (about 93 million miles) separate us, we could not be more connected. Literally, everything you see comes from the Sun. The words you are reading now are really photons that left the Sun a little over eight minutes ago only to bounce off this page and into your eyes. We owe our very existence to our Sun. It provides just enough heat to keep our fragile bodies from freezing to ice or burning to a crisp. Every bite of food we eat we owe to the Sun, whose energy is converted into plants that provide sustenance for everything up the food chain.

We may not have understood what it was or how it worked, but for millennia humanity has understood the Sun's importance. The earliest humans, awestruck by its blazing splendor, left drawings of this closest of stars on cave walls. Nearly every civilization, no matter where it sprang up on the planet, has revered the Sun. Myths about the Sun were the basis of the earliest deities of ancient Mesopotamian, Hindu, Egyptian, Chinese, and Mesoamerican cultures. Before Apollo, the ancient Greeks worshiped the Sun god Helios. Before Jupiter, the ancient Romans worshiped Sol.

On February 24, 2011, a solar flare blew out a huge waving mass of plasma.

Throughout our history, the Sun has been central to humanity's quest for meaning in the universe. But our roughly 100,000-year history has been a brief moment in our Sun's 4.5-billion-year life. Only recently, through advances in science and technology, have we begun to really understand our Sun—where it came from, how it functions, how it affects our lives, and how it eventually will destroy our planet.

This book was not written for scientists or for anyone with more than a rudimentary understanding of astronomy. For that matter, it was not written by a scientist either. It was written by an expert in energy for an audience with a passing interest in and an intractable wonder at our closest star. Like many of you, my earliest memories of the Sun involve painful burns after too much exposure. Beyond that, my knowledge of the Sun consisted of no more than what I learned from grade-school science textbooks.

It was my fascination with energy later in life that sparked an intense curiosity in the Sun. I wanted to understand everything I could about energy: where it came from, how it traveled, how it was used, and how we could use it more efficiently. There is simply no way to understand energy without understanding our Sun, the battery that powers nearly everything in the solar system. But I was no scientist. To understand the science of the Sun, I had to break it down into terms a nonscientist could comprehend. It was my pursuit of simplicity that was the genesis of this book.

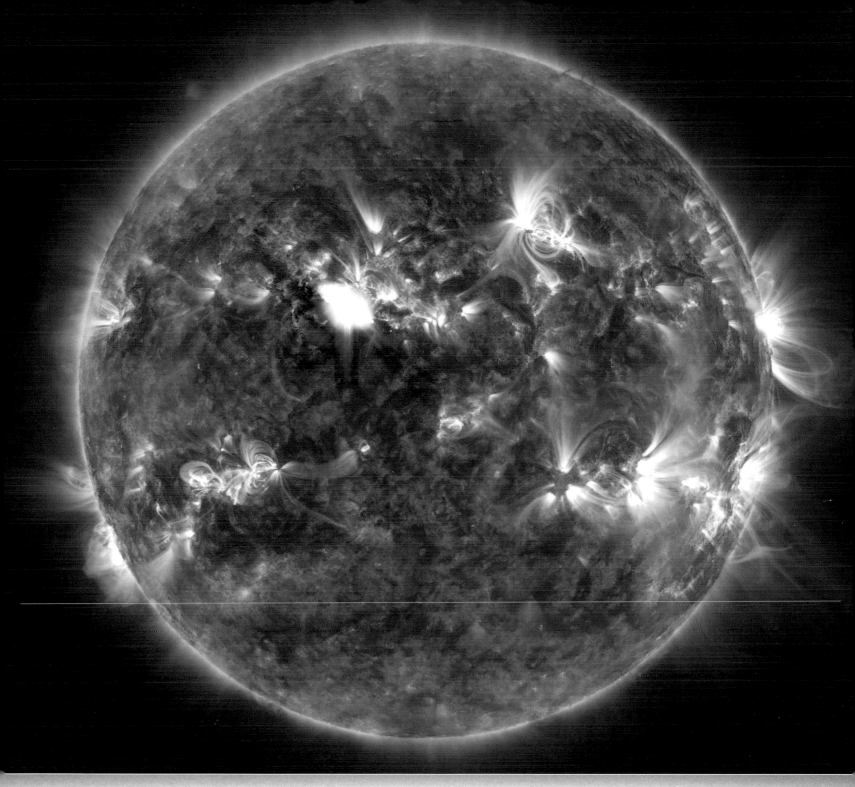

Magnetic loops are clearly visible in this image taken by NASA's Solar Dynamics Observatory on April 11, 2013. The bright spot is a midlevel solar flare.

But the science merely sparked an idea. It was the pictures that drove an obsession. In February 2010, the National Aeronautics and Space Administration (NASA) launched the Solar Dynamics Observatory (SDO). Soon afterward, it began producing images of the Sun unlike anything anyone had ever seen before (certainly unlike anything I had ever seen). Words cannot do them justice, which is why the pages that follow are strewn with some of the most stunning images. It is hard to look at these pictures and not become fascinated with our Sun. Though they have taught us much about the science of the Sun, these images transcend science. They are works of art that move the soul in the way only the best art can. They approach the spiritual, the sublime. They remind us not only of our physical connection to this immense ball of fire, but of the role the Sun has played in the shared development of all human culture. Everyone on the planet (and everyone who has ever *been* on the planet) has experienced our Sun. And, whether you realize it or not, like me, you have been profoundly changed by the experience.

An engraving of a 1635 illustration of the surface of the Sun by German astronomer Christoph Scheiner

Just as dramatic as the images, however, is our Sun's biography, the story of its birth and development, and the role it plays in a cosmic environment whose nature is only starting to be revealed. When reading the pages that follow, perhaps you, too, will wonder anew at the complex and dynamic relationship our Sun has with humanity and meet again—or perhaps for the first time—the famous (often quirky) individuals who have played important roles in helping us understand our Sun's life.

You will learn about our Sun's past, tracing its genealogy all the way back to the very first seconds of the Big Bang. You will read about the Sun's ambitious parent, how it likely has long-lost siblings still wandering the galaxy—and how some scientists believe one of these siblings may have stuck around, providing our Sun with constant (if hidden) companionship.

You will discover what makes our Sun tick, how and why it glows and spins. You will learn how scientists recently cracked the mystery of why the Sun's outermost layer is so much hotter than its inner layers. You will understand the science behind nuclear fusion (without needing an advanced degree) and why sunspots, solar flares, and coronal mass ejections form. More importantly, you will understand how all three can dramatically impact your life as well as gain some insight into why solar activity is important to global climate change. Along the way, you'll read about the geopolitical fight over who gets credit for inventing the telescope (and who *should*).

You will read about some of the bizarre mythologies surrounding Sun worship in cultures throughout the world. Yet you may be surprised to learn that some of the ancients had a better understanding of our Sun and how the universe works than most people did up until the eighteenth century. You will understand the true nature of light, how we evolved the ability to see, and the miracle (or, for some, the curse) of photosynthesis.

You will be forced to grapple with the fact that the Sun is trying to kill you even as it provides essential nutrients to keep you alive. You will hear of the new ways we are harnessing the Sun's energy and gain a better understanding of the old ways. You will read about the risks our Sun poses for modern electronics and learn some helpful tips on how to protect yours.

Finally, you will get a good picture of our Sun's future, including why humanity must figure out a practical mechanism for interstellar travel if it has any hope of surviving. You will also get a glimpse of what NASA has planned for the future of solar exploration and a renewed appreciation for the benefits this consistently underfunded agency brings to people around the world.

A modern rendering of the Aztec calendar stone with the Sun god Tonatiuh in the center. The original stone, found buried beneath Mexico City's main square in 1790, is about 3.7 meters (12 feet) in diameter and weighs more than 24 tons.

SOLAR EXPLORATION

We are fortunate to be living in a time of exciting new discoveries about our Sun. We are beginning to answer important questions about the structure of the Sun, questions that have perplexed scientists for centuries. Our expanding understanding of our Sun is due in no small part to investments that NASA made as early as the 1990s to investigate how solar activity impacts the rest of the solar system. In its 2007 Strategic Plan, NASA proposed a series of modest-sized solar missions that would deploy in fairly rapid succession to form a fleet of spacecraft working in tandem to monitor and analyze solar activity. Instead of a single, expensive mission to the Sun, NASA planned to launch 12 spacecraft, collectively known as the Heliophysics Great Observatory (HGO), positioned around the Sun, the Earth, and at strategic points between the two. Operating the spacecraft as a single, integrated observatory will provide multiple measurements of the same event and help fill observational gaps that occur when satellites monitor events from a fixed perspective. The most important of these spacecraft are:

SOLAR DYNAMICS OBSERVATORY

On February 11, 2010, NASA launched the flagship of the HGO, the Solar Dynamics Observatory (SDO). SDO was designed to spend five years investigating how our Sun generates its magnetic field, how the field is structured, and how solar magnetic energy is released into space. In the course of its investigation, SDO employs a suite of instruments designed to measure ultraviolet (UV) radiation, map solar magnetic fields, capture variations in the Sun's structure over the course of a solar cycle, and, importantly, to take high-resolution images of the Sun's surface and atmosphere.

Few were prepared for the jaw-dropping images SDO captured. With its Atmospheric Imaging Assembly, SDO takes images of the Sun in 10 wavelengths every 10 seconds (one wavelength each second). Not only has the Assembly resulted in a film-like image series of dramatic disturbances on the Sun's surface, it has produced gorgeous multicolored images worthy of framing.

The images in this book speak for themselves. SDO captures the Sun at twice the resolution of any other solar imager and at four times the resolution that NASA could produce in the late 1990s. This resolution has already helped NASA make exciting discoveries of solar events never before seen and provided visual confirmation of events that were once merely theoretical.

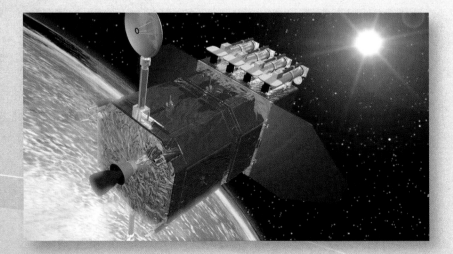

Left: An artist's rendition of NASA's Solar Dynamics Observatory

Opposite page: The large dark area in the Sun's atmosphere is a coronal hole, where the Sun's magnetic field opens up and allows the fast solar wind to escape.

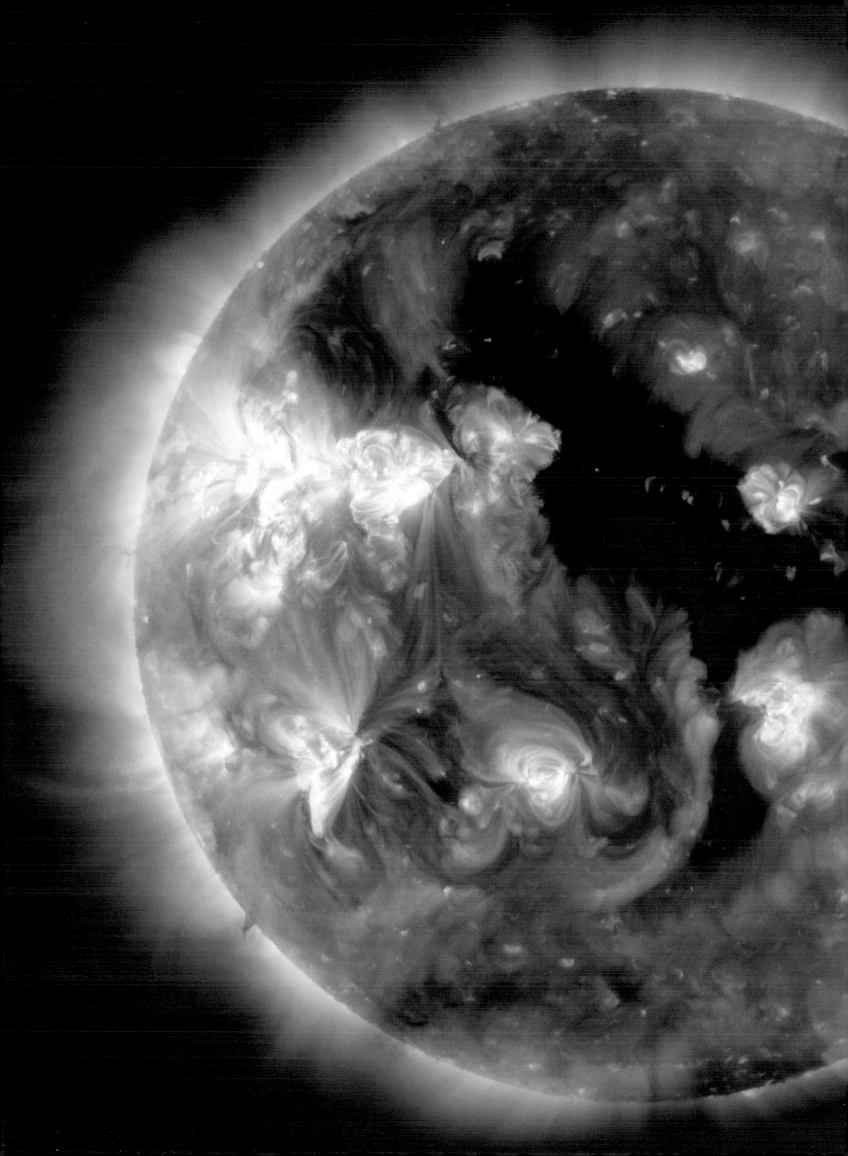

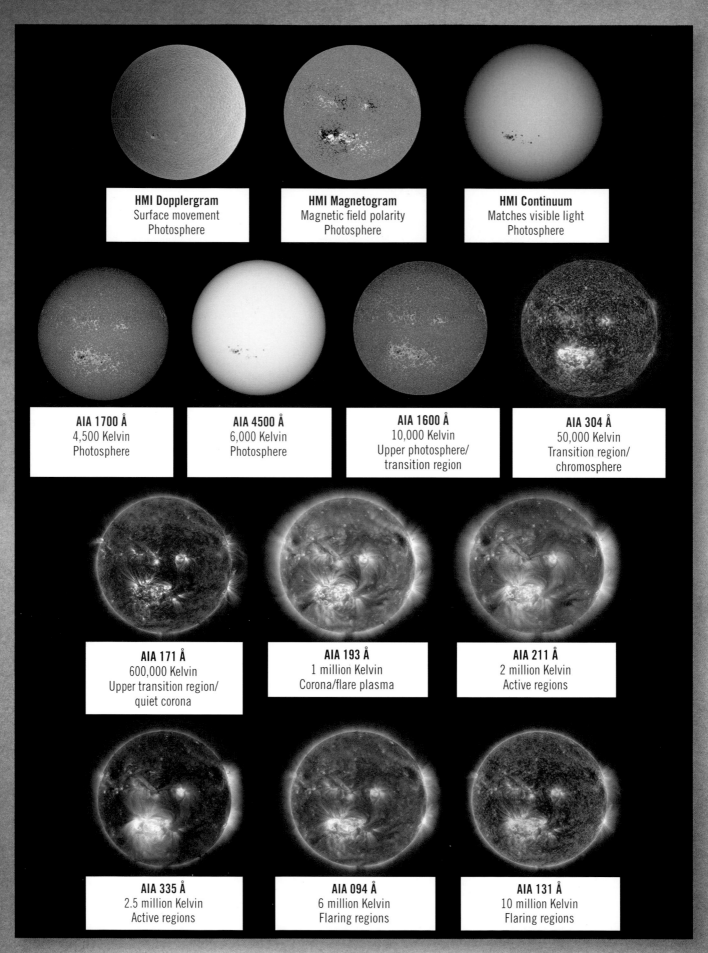

HMI Dopplergram
Surface movement
Photosphere

HMI Magnetogram
Magnetic field polarity
Photosphere

HMI Continuum
Matches visible light
Photosphere

AIA 1700 Å
4,500 Kelvin
Photosphere

AIA 4500 Å
6,000 Kelvin
Photosphere

AIA 1600 Å
10,000 Kelvin
Upper photosphere/
transition region

AIA 304 Å
50,000 Kelvin
Transition region/
chromosphere

AIA 171 Å
600,000 Kelvin
Upper transition region/
quiet corona

AIA 193 Å
1 million Kelvin
Corona/flare plasma

AIA 211 Å
2 million Kelvin
Active regions

AIA 335 Å
2.5 million Kelvin
Active regions

AIA 094 Å
6 million Kelvin
Flaring regions

AIA 131 Å
10 million Kelvin
Flaring regions

The wavelengths observed by NASA's Solar Dynamics Observatory (SDO) emphasize specific aspects of the Sun's surface or atmosphere. The imagery here is from the Helioseismic and Magnetic Imager (HMI), which focuses on the movement and magnetic properties of the Sun's surface, and the Advanced Imaging Assembly (AIA), which shows how solar material moves around the Sun's atmosphere.

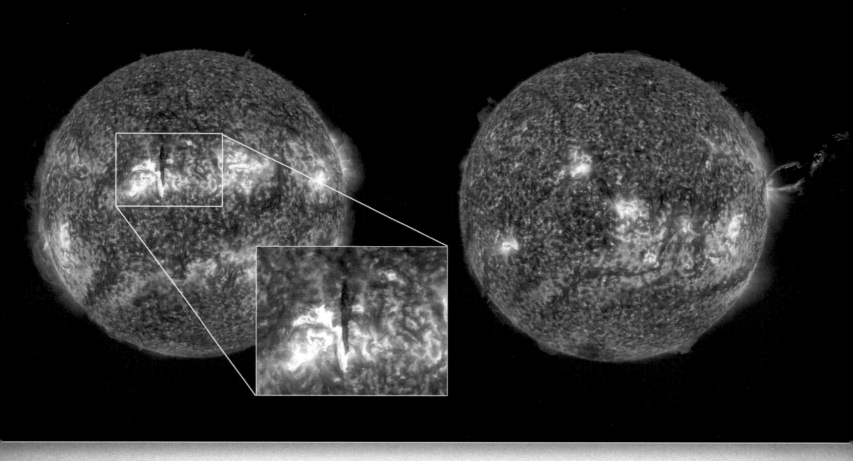

The image on the left, taken by STEREO-B (behind), shows a dark vertical line that the image on the right, captured by STEREO-A (ahead), reveals as a giant prominence bursting through the Sun's atmosphere.

SOLAR TERRESTRIAL RELATIONS OBSERVATORY

On October 26, 2006, NASA launched the Solar Terrestrial Relations Observatory (STEREO), two nearly identical spacecraft that orbit the Sun on nearly the same path as the Earth. While one pulled ahead of Earth's orbit, the other fell behind, providing stereoscopic (three-dimensional) images of the Sun. As their orbits gradually separated, STEREO could monitor the far side of the Sun. For the first time, NASA was capable of near real-time observation of the Sun from all angles. This allowed much more accurate detection of sunspots and more timely alerts regarding coronal mass ejections that may originate on the far side. The STEREO spacecraft were designed with no mission end-time. In theory, they could continue to monitor the Sun indefinitely.

An artist's rendition of NASA's STEREO spacecraft in orbit around Earth

SOLAR AND HELIOSPHERIC OBSERVATORY

The Solar and Heliospheric Observatory (SOHO) was launched on December 2, 1995, as a joint project between NASA and the European Space Agency (ESA) to study the Sun over two years. The spacecraft was placed into an orbit near the Sun-Earth L_1 Point, the point where the Sun and the Earth have an equal gravitational pull on the craft. This orbit enables SOHO to remain in a fixed position relative to both the Sun and the Earth. On board it carries 12 different instruments that can monitor the Sun independently or work in any combination for coordinated observations. Initially, SOHO was intended to investigate the Sun's outer layers, probe its interior structure, and observe the solar wind. However, it has remained in active use, becoming the main source of data for predicting solar weather and alerting Earth to potential hazards coming from the Sun.

In this composite image of a fiery coronal mass ejection from NASA's Solar and Heliospheric Observatory, a separate image of the Sun was enlarged and superimposed over the occulting disk for a dramatic effect.

REUVEN RAMATY HIGH ENERGY SOLAR SPECTROGRAPHIC IMAGER

NASA launched the Reuven Ramaty High Energy Solar Spectrographic Imager (RHESSI) on February 5, 2002, with the mission of studying the causes and impacts of solar flares. The spacecraft contains custom-made instruments designed to capture images of X-rays and gamma rays emitted by solar flares. RHESSI has not disappointed. It was the first satellite to capture images of gamma rays emitted by the Sun. On December 6, 2002, RHESSI also became the first satellite to image the electromagnetic radiation from a gamma-ray burst, the most energetic type of explosion in the universe. While observing the Sun, RHESSI caught a glimpse of a burst originating from far outer space and managed to measure its magnetic polarity.

With these remarkable spacecraft and the instruments they carry (and with newer craft set to launch over the next few years), we are beginning to unlock the Sun's secrets and reveal the close connections Earth has with our closest star. These solar missions are transforming our view of the solar system. As we learn more about the Sun, we are discovering that the solar system is far more than a set of objects floating in space, bound together by the force of gravity. What is emerging is an entire cosmic ecosystem, where more than gravity is at work. Material and energy is shared as forces overlap. Changes in one area may have profound effects on others. More than ever before, we are beginning to appreciate the Sun's place in the solar system, less for its enormous size and intense energy, and more for the complex relationship it has with Earth's geology, atmosphere, and, perhaps most importantly, its people.

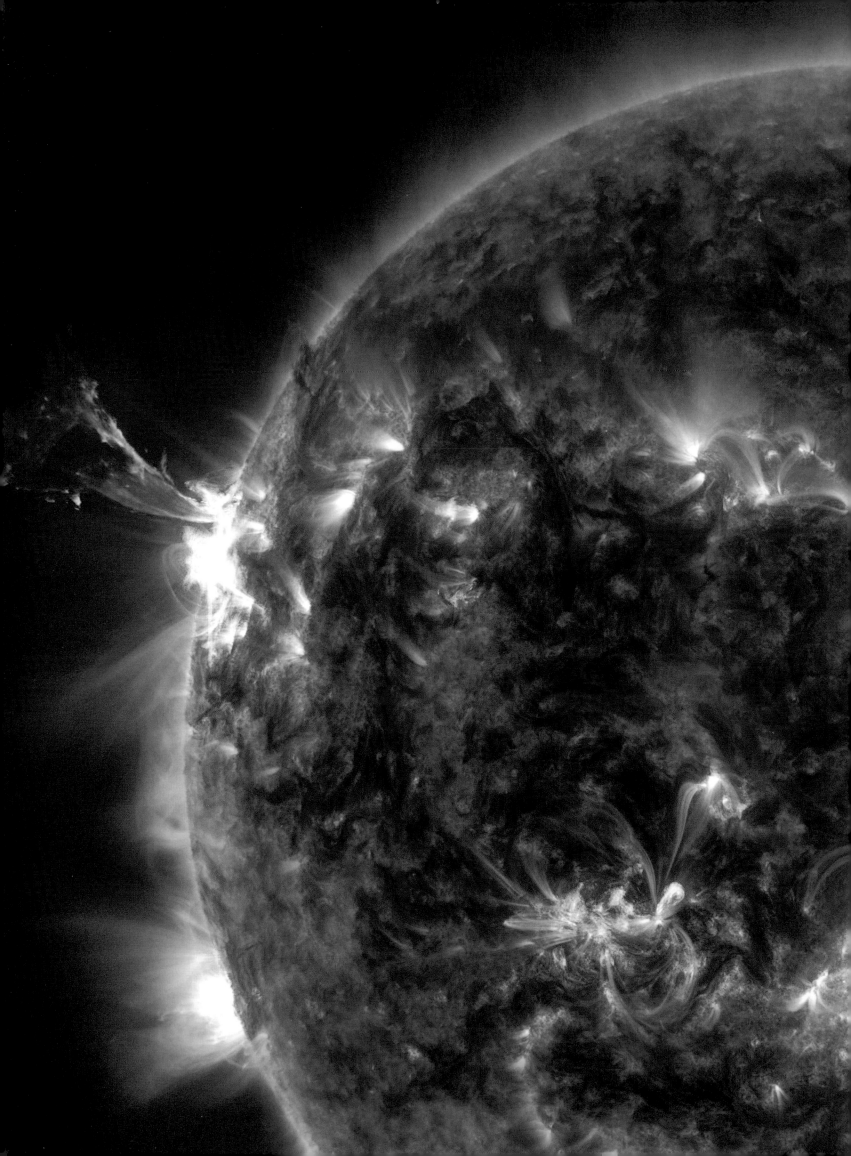

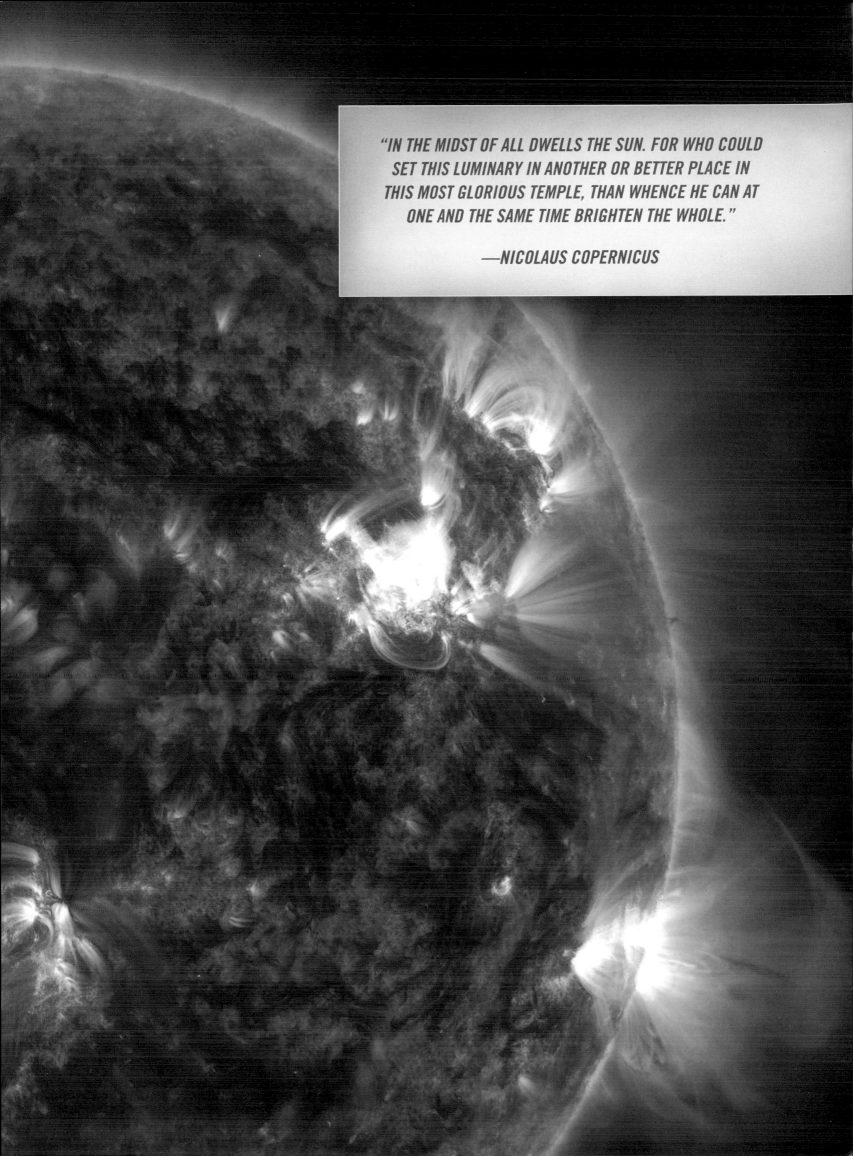

"IN THE MIDST OF ALL DWELLS THE SUN. FOR WHO COULD SET THIS LUMINARY IN ANOTHER OR BETTER PLACE IN THIS MOST GLORIOUS TEMPLE, THAN WHENCE HE CAN AT ONE AND THE SAME TIME BRIGHTEN THE WHOLE."

—NICOLAUS COPERNICUS

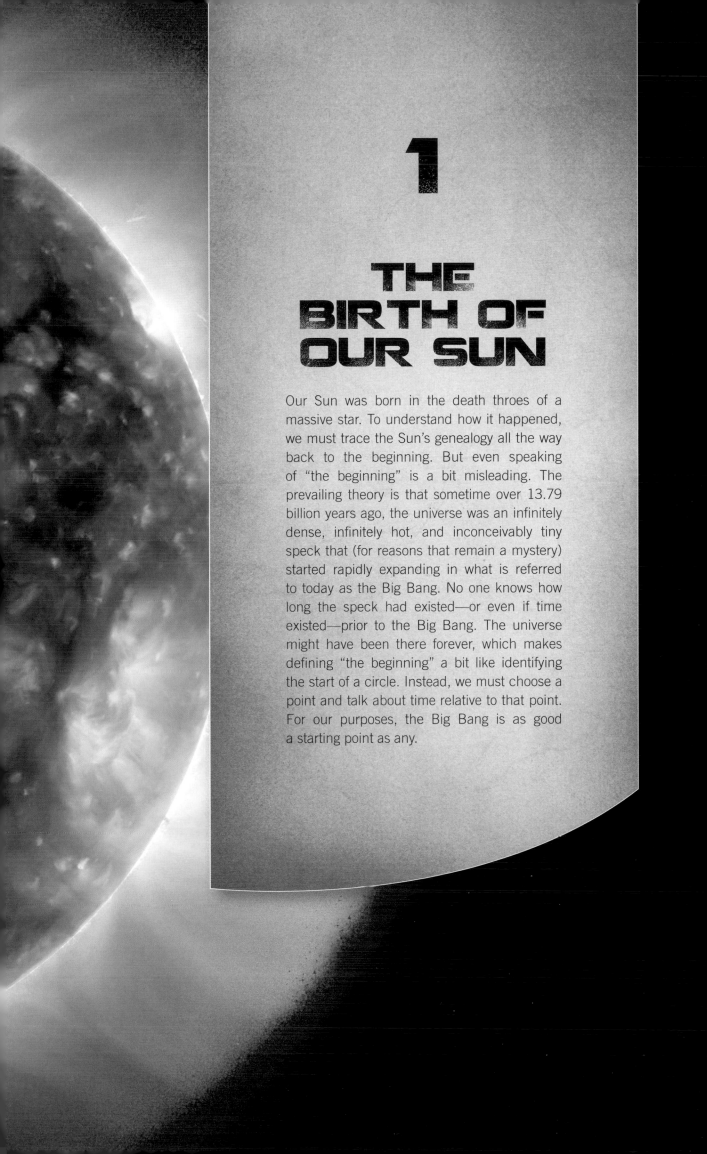

1

THE BIRTH OF OUR SUN

Our Sun was born in the death throes of a massive star. To understand how it happened, we must trace the Sun's genealogy all the way back to the beginning. But even speaking of "the beginning" is a bit misleading. The prevailing theory is that sometime over 13.79 billion years ago, the universe was an infinitely dense, infinitely hot, and inconceivably tiny speck that (for reasons that remain a mystery) started rapidly expanding in what is referred to today as the Big Bang. No one knows how long the speck had existed—or even if time existed—prior to the Big Bang. The universe might have been there forever, which makes defining "the beginning" a bit like identifying the start of a circle. Instead, we must choose a point and talk about time relative to that point. For our purposes, the Big Bang is as good a starting point as any.

THE ANTIMATTER UNIVERSE

Had antimatter gotten the upper hand during the earliest moments in time, no one knows what the universe would have looked like. To help give us an idea, scientists placed a seven-ton particle detector called the Alpha Magnetic Spectrometer (AMS-02) on the International Space Station. They are already processing data that should give us a better picture of the antimatter universe we could have become.

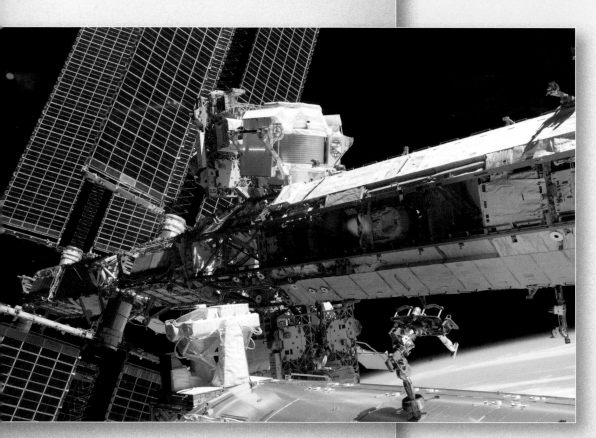

The AMS-02 installed atop the starboard truss of the International Space Station

THE BIG BANG

A fraction of a second—or .00000000000000000000 0000000000000000001 of a second, to be precise—after the universe came into existence, it started expanding at greater than light speed in a process physicists call cosmic inflation. That old adage that nothing can travel faster than light is not entirely true. Under rare circumstances—the earliest moments of the Big Bang, for instance—the laws of physics become more like strongly worded suggestions.

This normally impossible feat did not go on for long. In fact, cosmic inflation happened so quickly that describing it in fractions of a second would require more zeros than could practicably be printed on this page. By the end, the universe and everything in it—including the seeds of life—expanded to about the size of a grapefruit. While that might make the *Big* in Big Bang sound a bit exaggerated, keep in mind that in a tiny fraction of a second, the universe expanded from an infinitely small speck to something you could hold in your hand. From infinitely small to anything is quite a jump.

Just after cosmic inflation, the universe consisted of an extremely hot stew of subatomic particles. The stew was so hot that the energy that went into creating anything other than energy resulted in pairs of matter and antimatter. Because matter and antimatter get along like the Capulets and Montagues, the early universe was very unstable. As soon as tiny bits of matter and antimatter burst into creation, they collided with one another and were destroyed.

All of these fits and starts kept the early universe in a constantly shifting state until (for some unknown reason) matter started winning some of the clashes. The result was that, after each collision, rather than matter and antimatter destroying each other, a tiny excess of matter remained. If this sounds impossible based on everything you know about physics, you're right. It shouldn't happen. But through a reaction called baryogenesis, it did. As more and more collisions occurred, matter started to dominate antimatter, and the universe as we know it began to take form.

Eventually (and, again, we are talking tiny fractions of a second), the universe slowed down and cooled. Temperatures fell below the point where everything created came out as matter and antimatter pairs. From this point on, the density of the universe was dominated by matter—protons, electrons, neutrinos, and all the other bits that are the stuff of high-school physics classes.

After about 378,000 years, this cooling allowed protons and electrons to pair up, forming simple, neutrally charged hydrogen atoms (one positively charged proton balanced by one negatively charged electron). This point, an event known majestically as the Era of Recombination, marks the first time light appeared in the universe. Prior to electron-proton pairing, light energy never got anywhere since it was scattered by constantly colliding with all the free-floating particles in the chaos of the hot, early universe. Once light could travel without being scattered, for the first time in . . . well . . . *time*, the universe burst into view.

Omega Centauri is a massive globular cluster that orbits the Milky Way galaxy. It boasts nearly 10 million stars, which range between 10 and 12 billion years old. In this detail, the yellow-white stars are adult stars, like our Sun. Orange indicates late-life stars that have become cooler and larger, red indicates stars that have become red giants, and blue indicates stars that are fusing helium in their superhot cores in a desperate attempt to extend their lives.

THE EARLY UNIVERSE IN MICROWAVES

The Era of Recombination is the earliest we can peer back into history and "see" anything in the form of light still bouncing about. The light waves emitted at that time have since stretched to longer and longer wavelengths, which are now so long they exist as microwaves. These microwaves are what NASA's Wilkinson Microwave Anisotropy Probe (WMAP) detected when it looked as far back as anything could and created an anisotropy map of the early universe. *Anisotropy* is an intimidating word that simply means "not uniform." It refers to the interesting variations in temperature WMAP detected in the very first stages of the universe. In 2006, John Mather of NASA's Goddard Space Flight Center and Professor George Smooth of the University of California at Berkeley won the Nobel Prize for the insight these very small temperature variations provided to our understanding of how the first galaxies formed.

Created from WMAP data, this image of the universe at just over 370,000 years old reveals tiny temperature fluctuations (shown as color differences) that correspond to the seeds that grew into galaxies.

STAR NURSERIES

Matter in the early universe was distributed almost, but not quite, uniformly. Thanks to the tiny variations detected by WMAP, over a very long period of time, regions of slightly greater density exhibited slightly more gravitational pull. As these regions attracted more and more matter, they became even denser, pulling in even more matter and forming vast clouds known as nebulae.

Just as tiny variations in the density of the universe created nebulae, tiny variations in density within the first nebulae made the giant dust clouds gravitationally unstable. At denser points within the nebulae, matter began to clump together. When things clump together, their contraction causes the clumps to spin—a phenomenon the astronauts observed while floating weightlessly about in their spacecraft. The faster something spins, the more the matter toward the outside of the clump flattens out. This flattening effect is partly why most galaxies take on a disk-like shape.

After about 100,000 years, all of this spinning, flattening, and contracting released enough energy that the clumps of matter began to heat up, glow, and create pressure in the form of infrared radiation pushing back against the force of gravity. When contracting matter reaches the point where it begins to emit infrared energy, it is called a protostar.

Over another 100,000 years or so, these protostars continued to increase in density, attract more matter, and grow even hotter. Eventually, the high temperature and massive pressure at their centers began forcing hydrogen atoms to fuse together, sparking a fusion reaction. Astronomers refer to this reaction imprecisely as hydrogen burning. When the fusion process begins, a star is born. A new star is sometimes referred to as a *nova*, Latin for "new."

With its thousands of sparkling young stars, NGC 3603 is among the most massive young star clusters in the Milky Way Galaxy.

WHEN GRAVITY DEFEATS FUSION

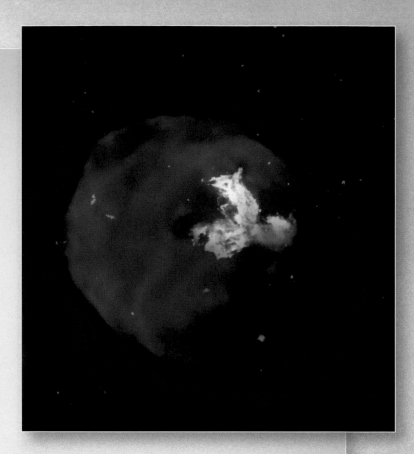

A star can turn supernova in one of two ways: thermal runaway or core collapse. In thermal runaway, a star accumulates so much mass that gravity raises the temperature of its core high enough to ignite a runaway nuclear reaction. During normal fusion, hydrogen atoms fuse together, forming helium. If temperatures increase enough, the helium atoms fuse, forming carbon. In stars with mass about the size of our Sun, this is where the fusion cycle stops. Their core temperatures will never become hot enough to ignite carbon fusion, and they will continue to burn through their remaining helium until they form a white dwarf composed mostly of cold, inert carbon.

However, the cores of some stars with sufficient mass can reach about 15 million degrees Celsius (more than 27 million degrees Fahrenheit), the temperature at which carbon atoms fuse together. Once the temperature inside a star reaches the point of carbon ignition, there is no stopping the fusion cycle. Each successive fusion process converts more of the star's remaining mass, while producing temperatures hot enough to fuse together the resulting elements. Consequently, each fusion conversion burns faster and faster.

It can take billions of years to fuse all of a star's hydrogen into helium. It can take millions of years to fuse its helium into carbon. Carbon fusion, on the other hand, can happen in hundreds of years, a relative instant in cosmic terms. When carbon fuses together, it creates neon. Neon fusion can convert a star's entire carbon stocks within a matter of years. Next is oxygen fusion, which can take as little as a couple of months.

The succession of cycles becomes so fast, with each cycle converting more mass while releasing less energy, that at some point, the outward pressure of fusion loses its battle with the inward pressure of gravity—and the balance of energy that held together the entire structure suddenly collapses. All of the gravitational force that was once held in check comes crashing inward. Gravity crushes subatomic particles together. This instant collapse of the atomic structure of every atom within the star's core releases a shock wave of energy that rips through the star, blasting most of its remaining mass into space at supersonic speeds.

In thermal runaway, the outward pressure of successive fusion cycles grows weaker until it is overcome by the force of gravity. But gravity sometimes can overcome outward pressure no matter what type of fusion is happening in the star's core. During core collapse, the core of very massive stars may begin to exert so much gravitational pull that even the energy created through hydrogen fusion cannot hold the structure together. Rather than wait for fusion to weaken, gravity skips straight to crushing together subatomic particles in the core. The result is the same shockwave that bursts the star from the inside out.

Cassiopeia A is the remnant of a supernova that blew the star apart about 11,000 years ago.

GOING SUPERNOVA

Once nuclear fusion is sparked, you can think of a star as existing in a constant tug-of-war. On one side is the inward force of gravity, constantly trying to compress the star into a denser and denser mass. On the other side is the outward force of fusion, creating pressure in the form of energy wanting to escape out into the universe. As long as these two forces stay in relative balance, the star lives a long and sunny life, steadfastly bringing light and warmth to its little corner of the universe. But if a star attracts enough mass, gravity will tug harder than fusion, collapsing the star's internal structure and sparking a supernova.

A supernova is among the most violent events in the universe. In a single flash, a supernova can release as much energy as an average star will emit during its entire lifetime. This is an enormous amount of energy. The atomic bombs that the United States dropped on Hiroshima and Nagasaki during World War II released energy equivalent to about 15 kilotons of TNT. Since then, the U.S. has developed hydrogen bombs with maximum yields of 25 megatons, nearly 2,000 times stronger. A single supernova releases the same amount of energy as about 10 trillion of these 25-megaton nuclear bombs. Even in cosmic terms, supernovas are a big deal. When they occur—about every 50 years in the Milky Way—they are, for a short time, the single brightest objects in the entire galaxy.

The Crab Nebula is the result of a supernova observed by astronomers in 1054. At its center is a pulsar, a neutron star that, despite only being the size of a small town, has the mass of our Sun.

SUPERNOVA NUCLEOSYNTHESIS

Stars that spark supernovas as a result of thermal runaway can, for a short time, fuse carbon into oxygen and then oxygen into silicon. For about a day, these stars fuse silicon into iron. But silicon fusion is as far as it goes. Fusing iron or any heavier element absorbs more energy than it produces. As a result, they can never be used to fuel a star.

During oxygen and silicon fusion, a massive star can also create sulfur, chlorine, argon, scandium, and titanium. The enormous temperatures and pressures created during a supernova also produce the essential nutrients sodium, potassium, and calcium. When famed astronomer Carl Sagan proclaimed, "We're made of star stuff," he was right. You are alive and reading these words because you have within you elements that were first created in a massive supernova explosion that occurred billions of years ago. Stop for a second and think about that. You are composed of stuff that can only be created in one of the most violent explosions in the universe. Inspiring . . . and somewhat frightening.

Fortunately, massive stars that are destined for a violent end have an upside. Most stars do not contain enough mass to spark a supernova. Instead, they spend their lives fusing hydrogen into helium and helium into carbon until the fusion reaction in their core simply peters out. If the universe were made only of these kinds of stars, space would contain mostly (if not entirely) hydrogen, helium, and carbon, a very limited set of blocks on which to build the panoply of life. All of the heavier elements in the universe are created only by the intense heat and energy of a supernova in a process known as supernova nucleosynthesis.

Within a few seconds of a supernova explosion, so much energy is released that, for a brief time, the explosion generates temperatures much higher than those found in the core of any star. These superhigh temperatures create all of the heavier elements from cobalt to uranium, as well as their various isotopes. Isotopes are versions of the same element that differ by the number of neutrons in their nuclei. Some isotopes—called radioisotopes—decay at a steady rate as they release excess neutrons over time.

HYPERNOVAS AND GAMMA-RAY BURSTS

Recently, scientists have classified a specific type of core collapse called a hypernova. Usually, a supernova is characterized by a star exploding as a result of thermal runaway or core collapse. But for stars with masses about 15 times that of our Sun, the outward burst of energy created by collapsing matter in the star's core is insufficient to blow the outer layers beyond the pull of gravity. Instead of exploding heavy elements into the universe, most of the star's mass collapses in on itself, forming a black hole. Only some of the mass is able to escape as jets of high-energy particles called gamma-ray bursts. These bursts are the brightest electromagnetic events known to occur in the universe.

Because they are usually short-lived, focused beams, gamma-ray bursts are extremely powerful. So far, astronomers have only observed them (about one a day) emanating from distant galaxies. If a gamma-ray burst were to occur in the Milky Way and the Earth was in its path, the results would be catastrophic. The planet would be bathed in so much ultraviolet radiation that even the Earth's protective magnetosphere could not prevent most organisms from being killed. In fact, some scientists speculate that exposure to a gamma-ray burst is precisely what triggered the Ordovician-Silurian mass extinction event some 450 million years ago.

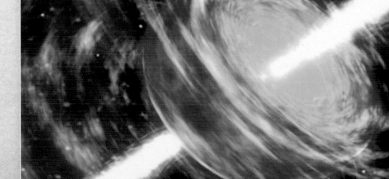

A brilliant gamma-ray burst jets out from the center of a dying star in this computer image of a hypernova.

Also known as the Tarantula Nebula, 30 Doradus is a huge star-forming region in a galaxy called the Large Magellanic Cloud. In the center, thousands of massive stars are blowing off material and producing intense radiation and powerful winds.

OUR SUN'S VIOLENT CHILDHOOD

One particular radioisotope has helped scientists trace the origins of our Sun to a supernova explosion over 4.6 billion years ago. In the final stages of a massive star's life, some of the iron formed during silicon fusion captures an extra neutron and becomes iron-60, a radioisotope with a half-life of 1.5 million years. Iron-60 remains within the star's core and only escapes when it blows itself apart during a supernova explosion.

Because iron-60 releases radiation at a steady rate over a long period of time, scientists can use it as a kind of clock, a marker of supernova activity early in our Sun's formation. By knowing the approximate age of the Sun, they can estimate how much of the iron-60 that may have been created at the Sun's birth would have decayed by today. Then, they can look for the presence of iron-60 with that amount of decay.

This is precisely what some scientists did in the mid-2000s. Astronomers at Arizona State University theorized that bits of the radioisotope blown out from a supernova mixed with the gas and dust that were forming our Sun's protostar. But some of the material from the supernova must have become rocks floating about the solar system as it was forming. Eventually, the scientists theorized, some of those rocks must have fallen to Earth as meteors. So the scientists examined some of the meteorites that dated back to the birth of the solar system and found an abundance of iron-60 that had decayed exactly the amount one would expect if the protostar that created our Sun had been exposed to a supernova blast.

Based on these findings, the astronomers suggest that 4.6 billion years ago a massive star was born inside a giant dust cloud. Excess heated gas created a bubble that caused dust nearby to start forming a cluster of smaller protostars, among them the one that would become our Sun. When the massive star went supernova, it peppered the infant star cluster with iron-60 and ignited some of the protostars. According to this theory, the supernova both birthed our Sun and left its DNA to prove it.

THE IRON-60 CONTROVERSY

Recently, the supernova birth theory has come under attack by scientists at the University of Chicago. In 2012, a team of researchers there examined the same meteorites as the Arizona State University team, but used different methods to remove impurities that could cause errors in measuring radioisotope levels. After removing impurities, the Chicago team found much lower levels of iron-60 than the Arizona team had found, and no definitive evidence that a supernova was involved in our Sun's birth. The clashing results have sparked renewed speculation about our Sun's early development and a heated debate that may take several years and much more data to resolve.

An artist's rendition of a supernova blasting out gas and debris

FUSION: ENERGY OF COMBINATION

When the dust cluster that was our Sun's protostar reached a critical temperature, it ignited a fusion reaction that began converting hydrogen into helium. The moment this hydrogen fusion reaction began, our Sun became a star! It will spend 90% of its lifetime performing this conversion.

Just how long a star like our Sun—what scientists refer to as a main-sequence or dwarf star—will take to convert all of its hydrogen into helium depends on two factors: its mass and its luminosity. Its mass is a measure of how much fuel it has to burn, while its luminosity is a measure of how fast it burns that fuel.

A typical hydrogen atom contains a single electron orbiting a nucleus containing a single proton. There are eight known types of helium. But for the purpose of fusion within our Sun, we only need to understand three, which vary based on the number of neutrons, if any, in their nuclei. Helium-2 is called a diproton because its nucleus contains two protons but no neutrons. Helium-2 is unstable and tends to decay very quickly back into hydrogen. Helium-3 has a nucleus containing two protons and a single neutron. It is extremely rare on Earth, but thought to exist in large quantities in the surface layers of the Moon. Helium-4 has a nucleus with two protons and two neutrons. It is the most common type of helium on Earth. It is the stuff that makes balloons float and that kids love to inhale to make their voices sound high and squeaky.

PRESTO CHANGE-O!

You might be wondering how an atom with two protons can suddenly turn into an atom containing one proton and one neutron. How can one of its protons magically change into a neutron? It is rare, but possible. During beta-plus decay, a hydrogen atom ejects a positron (antimatter's version of an electron) and an electron neutrino.

Inside every proton and neutron are quarks, particles that have no known substructure. Quarks usually come in two types: up quarks or down quarks. The balance of up and down quarks within a larger subatomic particle will determine its charge.

When a proton emits a positron, some of the up quarks within it change to down quarks. This affects the charge of the proton, turning it from positive to neutral. Since a positron is antimatter, it immediately collides with (and annihilates) an electron. But the kinetic energy of both subatomic particles is emitted as gamma rays. The presence of these gamma rays is one way we know that the positron and electron once existed at all.

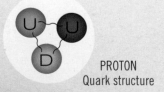
PROTON
Quark structure

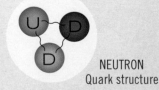
NEUTRON
Quark structure

The process our Sun (and other stars of its size) uses to fuse hydrogen into helium is called a proton-proton chain reaction. It is a multistep process that took physicists decades to work out. Here is a simplified version:

Step 1: Under immense temperature and pressure, two hydrogen nuclei (each with a single proton) are fused together, forming helium-2.

Step 2: Most of the time, helium-2 is so unstable that it quickly decays right back into hydrogen. But every now and then, helium-2 goes through a process called beta-plus decay, during which one of its protons changes into a neutron by emitting two subatomic particles: a positron and a neutrino. Helium-2, with one proton and one neutron, is called deuterium, or heavy hydrogen.

Step 3: The high temperature and pressure can fuse deuterium (with its one proton and one neutron) together with another proton, forming helium-3 (two protons with one neutron).

Step 4: Two helium-3 atoms (a total of four protons and two neutrons) are fused, forming a single helium-4 atom with a nucleus containing two protons and two neutrons. The by-product is two hydrogen atoms (each with a single proton nucleus).

Every time it occurs, this process releases about 13 megaelectron volts of energy. Since it takes about 624 billion megaelectron volts each second to power a 100-watt light bulb, 13 might not sound like much. But remember, that is the amount of energy generated by a process that consumes only four hydrogen atoms. And the Sun has a lot of hydrogen atoms—enough, in fact, to burn nothing but hydrogen for another 5.4 billion years. So even though each proton-proton fusion process releases relatively minute amounts of energy, so many processes are occurring at the same time throughout the Sun's enormous mass that the energy released at any given time is also enormous.

HYDROGEN FUSION IN THE SUN

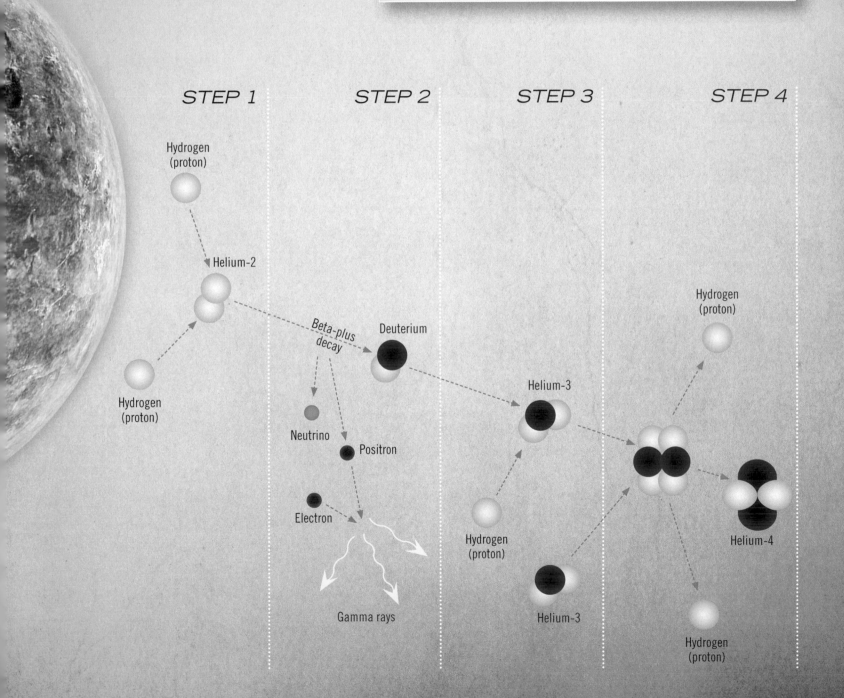

STEP 1   STEP 2   STEP 3   STEP 4

Hydrogen (proton)

Helium-2

Hydrogen (proton)

Beta-plus decay

Deuterium

Neutrino

Positron

Electron

Gamma rays

Helium-3

Hydrogen (proton)

Helium-3

Hydrogen (proton)

Hydrogen (proton)

Helium-4

Hydrogen (proton)

TWO'S COMPANY

A surprising number of stars have partners and exist in binary systems where the two stars orbit a central point at which their gravitational pull on each other cancels out. Until recently, large, bright stars were the most easily detected. As a result, astronomers mistakenly estimated that well over 60% of all the stars in our galaxy must exist in binary systems. Improved technology has helped modern astronomers discover that the vast majority of stars in our galaxy are cold, low-mass stars that have long run out of hydrogen fuel to burn. Only 3% to 4% of these types of stars exist in binary systems. So, taking these stars into consideration, binary systems are the exception, not the rule. However, about half the stars of average mass and luminosity—like our Sun—exist in binary systems.

Binary star systems can be detected using three different methods. Visual binaries, like the name implies, are systems that can be detected with the naked eye or with a telescope. Many times, however, brighter stars obscure visual detection of their partners, making them appear as singles. These binaries are usually detected through spectroscopy, a process that reveals the composition and relative movement of a star. Finally, astronomers can infer the existence of some binary systems from the awkwardness of a star's orbit. Some stars appear to orbit around empty space, with no visible companion star. Through precise measurements of the star's movements over a long period of time, astronomers can detect where an area of mass should be. These inferred star systems are known as astrometric binaries.

Like single-family homes, binary star systems usually take one of three forms: detached, semidetached, or contact (the townhouses of binary star systems). Most binary star systems are detached binaries, where the gravitational pull of each star has no meaningful effect on the material of the other. For all practical purposes, the two stars live separate lives. In semidetached binaries, one star's gravitational pull dominates the other's to the point that the dominant star sucks material from the surface of the subordinate star. These systems can sometimes be detected because the superheated gases stolen from the subordinate star form accretion disks, like the rings of Saturn, around the dominant star. In contact binaries, the material in each star is affected by the gravitational pull of the other. The stars share a common atmosphere and, under some conditions, may merge completely.

The greenish color indicates jets of gas escaping from "envelopes," the collapsing clumps of gas and dust that form stars. In the first two panels of the top row, twin stars have already formed inside the envelopes. Scientists believe the envelopes in other panels will also form binary stars.

OUR SUN'S LONG-LOST SIBLING?

If about half the stars the size and brightness of our Sun are binary systems, could our Sun have a companion that we simply haven't detected? Astronomers generally agree that our Sun was born among a cluster of stars with similar masses, temperatures, and luminosities. Most of them dismiss the idea that we can identify any of our Sun's long-lost siblings some 4.3 billion years after they left their stellar nursery and parted ways. But that hasn't stopped some astronomers from looking. Despite extensive searching, however, using powerful infrared telescopes capable of detecting red dwarfs as cool as 150 degrees Celsius (over 300 degrees Fahrenheit), no astronomical survey has found a companion star to our Sun.

In this artist's conception of a relatively young binary star system, the rocks and dust in the inner belt around the near star may be forming an Earth-like planet.

Still, there are astronomers who believe that the existence of a companion to our Sun can be inferred—like other astrometric binaries—from a number of astronomical anomalies. First, they point to the precession of equinoxes. If you watched the sky every night for a year, you might notice that the position of the stars slowly shifts across the sky. This movement is caused by the Earth's rotation around the Sun and the appearance of the stars from our shifting perspective. This is known as precession.

The precession of equinoxes is slightly more complex. Imagine that, instead of watching the sky every night, you took a single snapshot of the sky at the same time on the same night every year for 20 years. What you would notice is that, over time, the stars appear to shift backward. This is what's called the precession of equinoxes, or precessional movement. Astronomers have calculated that precessional movement follows a circular path that should, in theory, take about 26,000 years to come full circle.

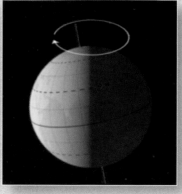

*Precessional movement is thought to be caused by the slow wobble of Earth on its axis.
One rotation takes about 26,000 years.*

Since you take your snapshot at the same time on the same night every year, precisional movement cannot be an illusion created by the Earth's orbit around the Sun. Something else must be causing the shift. Sir Isaac Newton was the first to postulate that the appearance of precessional movement is due to the Earth slowly wobbling on a shifting axis. Picture a rod running straight through the Earth from top to bottom. As the Earth spins around the rod (axis), the rod wobbles like a top, tracing out the shape of a cone at both poles. It takes roughly 26,000 years for the Earth to wobble fully around this circle. Most astronomers believe the wobble is the result of the interaction of lunar and solar tidal forces, and that it explains the precession of equinoxes.

But there's a problem. If the slow wobble of Earth's axis causes the precession of equinoxes, it is a product of our shifting perspective and should affect everything we view from Earth. Some astronomers argue that objects within our solar system do not appear to precess. Only objects outside of the solar system do. If this is the case, then the Earth's wobble cannot be the cause of precessional movement. Some astronomers argue that the fact that objects inside the solar system do not appear to precess is evidence of the existence of a companion star. If our Sun were part of a binary system, the whole solar system would be rotating around a common center of gravity with the companion star. A binary model, they claim, is the perfect explanation for why objects outside the solar system precess, and objects within the solar system do not.

Another argument used as evidence that our Sun has a companion involves long-period comets (comets with orbits that can take thousands of years). Although it has yet to be observed, astronomers believe that the outer edge of our solar system is surrounded by a dense field of icy rocks called the Oort Cloud. It is thought to be the detritus of our solar system's early formation, flung outward by the gravitational forces of the gas-giant planets—Jupiter, Saturn, Uranus, and Neptune. Most astronomers believe that long-period comets originate in the Oort Cloud. Every so often, the gravitational force of an object passing outside of the Oort Cloud will disturb icy chunks within it and send them hurtling toward the inner solar system. Some of these objects are swallowed by the Sun. But sometimes their trajectory causes them to be caught by the Sun's gravitational force only enough to be slung around the star and shot back out into the solar system. This gives the object—now a comet—an elongated orbit that it will follow for millions of years (assuming it doesn't hit anything).

Some astronomers claim that the distribution of these long-period comets is not random. The comets appear to originate in a specific location in the Oort Cloud and are flung at similar trajectories into long-period orbits. For these astronomers, this is evidence of the existence of something with a significant gravitational force that disturbs the Oort Cloud on a cyclical basis. Many claim that a solar companion that orbited about every 26,000 years would explain this regular distribution of long-period comets.

In the mid-1980s, two different teams of astronomers published academic papers speculating that a hidden companion star, likely a white dwarf, orbited beyond the Oort Cloud, approximately 1.5 light-years from the Sun. They called this hypothetical star Nemesis, or the Death Star, because its orbit was thought to coincide with what appeared to be a cycle of mass extinctions in Earth's geological record. According to their calculations, galactic forces and the gravitational impact of passing stars would give Nemesis an irregular, sharply elliptical orbit that disturbed the Oort Cloud in cycles matching the mass extinctions.

In 2011, Coryn Bailer-Jones, an astrophysicist at the Max Planck Institute for Astronomy, published a study in which he analyzed Earth's craters, looking for patterns that might explain how often large objects impact the planet. He found that evidence for a regular cycle of impacts was, in fact, a statistical error. His analysis indicated that, since about 250 million years ago, the rate at which objects impact the Earth has actually increased. Whatever was causing the increase, it certainly was not a Death Star lurking just outside the solar system. Bailer-Jones's paper poured some cold water on the growing parade of Nemesis enthusiasts. Nevertheless, die-hard believers still search for our Sun's long-lost sibling—and prepare for the doomsday its discovery is thought to foretell.

WHERE IS NEMESIS?

One reason we may not have found our Sun's long-lost sibling Nemesis (if it exists) is that it may be a cool, dim red dwarf star following a long elliptical orbit that brings it close to our solar system only every 26,000 years.

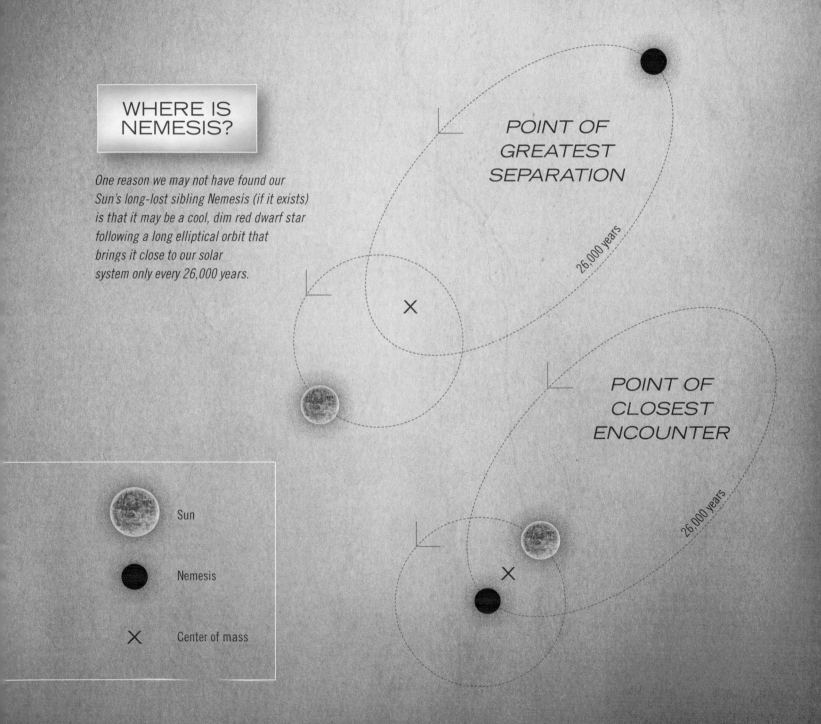

POINT OF GREATEST SEPARATION

26,000 years

POINT OF CLOSEST ENCOUNTER

26,000 years

Sun

Nemesis

X   Center of mass

THE STRUCTURE OF OUR SUN

As special as our Sun is to us, it is, by size and brightness, a rather average star. Stars like the Sun are known as G-type main-sequence stars. Don't be fooled, though. In star terms, average is still quite massive. Our Sun, for example, makes up 99.86% of all the mass in our solar system. Its diameter is about 1.4 million kilometers (roughly 870,000 miles). If that doesn't sound too impressive, picture this: if you lined up Earths, surface to surface, it would take over 109 of them to stretch the distance of the Sun's diameter. Put another way, even if you flew in the fastest plane ever constructed—NASA's unmanned X-43A, which can reach speeds of Mach 9.6 (11,265 kilometers, or about 7,000 miles, per hour)—it would still take you more than five days of continuous flying to travel through the Sun from one side to the other.

Helium
19.7%

Hydrogen
78.5%

Carbon
0.4%

Iron
0.14%

Oxygen
0.86%

Nitrogen, neon, sodium,
magnesium, aluminum,
silicon, phosphorus,
sulfur, potassium
0.54%

*About 78.5% of our Sun is made of hydrogen.
Most of the remaining mass is helium (19.7%),
the product of two hydrogen atoms fusing together.*

*All the other elements in our Sun
make up less than 2% of its mass.*

LAYERS OF OUR SUN

To talk about traveling from one "side" of our Sun to the other is a bit misleading. The Sun has no true surface, nothing hard on which to land an airplane (even one as advanced as the X-43A). In fact, the matter that makes up our Sun extends well beyond Earth, far out into the solar system. To be entirely accurate, therefore, everything in our solar system, including all eight planets, exists "inside" the Sun. Nevertheless, to make things easier, we can think of our Sun as having a central core surrounded by several layers. The inner layers terminate in a kind of outer crust, which is surrounded by a sort of misty atmosphere.

I write "a kind of" and "a sort of" because most of the Sun is made of plasma, a superheated form of matter that, like gas, does not retain a definite shape or volume that you can hold in your hands. On Earth, we think of matter as existing in one of three states—solid, liquid, or gas. In reality, most matter in the universe exists in a fourth state: plasma. When matter is superheated, its atoms go through a process known as ionization. In simple terms, ionization occurs when so much energy is applied to atoms that their electrons become excited enough to leap out of the atoms. Plasma is made up of free-floating, negatively charged, super-excited electrons

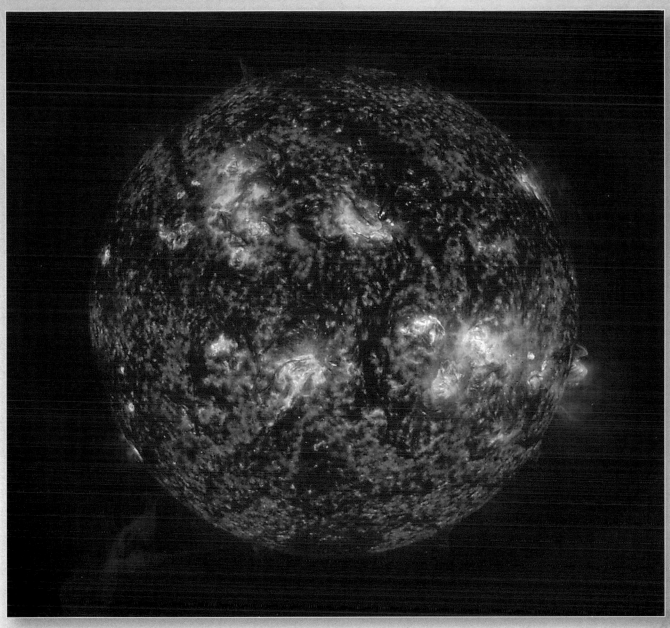

A long, twisted strand of plasma suspended in the Sun's corona

and the positively charged particles known as ions that remain after the electrons have leaped from the atom. Because ions are capable of carrying an electrical charge, plasma tends to be a strong electrical conductor. It reacts very easily to electromagnetic forces.

Our Sun is composed entirely of plasma, as are all stars. Most of the interplanetary space in our solar system contains plasma; so does most interstellar space. Even black holes, those voids so powerful that light cannot escape, excrete nothing but plasma.

Even though we speak of the Sun as having a surface and an atmosphere, the distinctions between the two are not quite so pronounced. The surface is not solid. It is made of the same plasma as in the Sun's atmosphere, only slightly denser. The lower-density plasma in the Sun's atmosphere makes its outer boundaries even more imprecise. Although we tend to think of the solar system as a collection of discrete objects (the Sun, the planets, the asteroids, and so on) separated by vast amounts of empty space, in reality the solar system is more like a giant plasma sea in which all of these objects float. In truth, the Sun's atmosphere extends to the outer edges of the solar system, its density slowly dissipating along the way. Knowing this, we can start to think of everything within our solar system as seamlessly connected and as existing, in a very real sense, *inside* the Sun.

THE CORE

The radius of the Sun's core is about 25% of the size of the entire Sun. The X-43A would still require over a day's flight to travel through it. But with an average temperature close to 15.7 million degrees Celsius (over 28 million degrees Fahrenheit), the high-tech plane would melt long before it completed the journey. The core is, by far, the hottest part of the Sun. In fact, it is the hottest place in our entire solar system. Although it accounts for only 10% of the Sun's volume, it is the source of almost all the Sun's heat.

Inside the core, gravity and pressure fuse hydrogen into helium, releasing vast amounts of energy that travel outward through each successive layer of the Sun. We tend to think of the Sun as an incredibly powerful inferno (and it is). But theoretical models predict that the power production density (the amount of energy produced by each unit of the Sun's volume) at the very center of our Sun is about the same as an active compost heap. The Sun's massive energy output is due not so much to its concentrated power as to its tremendous mass. Every second, the Sun converts 600 million tons of hydrogen to helium. Each individual conversion generates a tiny amount of energy. But there are a lot of individual conversions in 600 million tons every second!

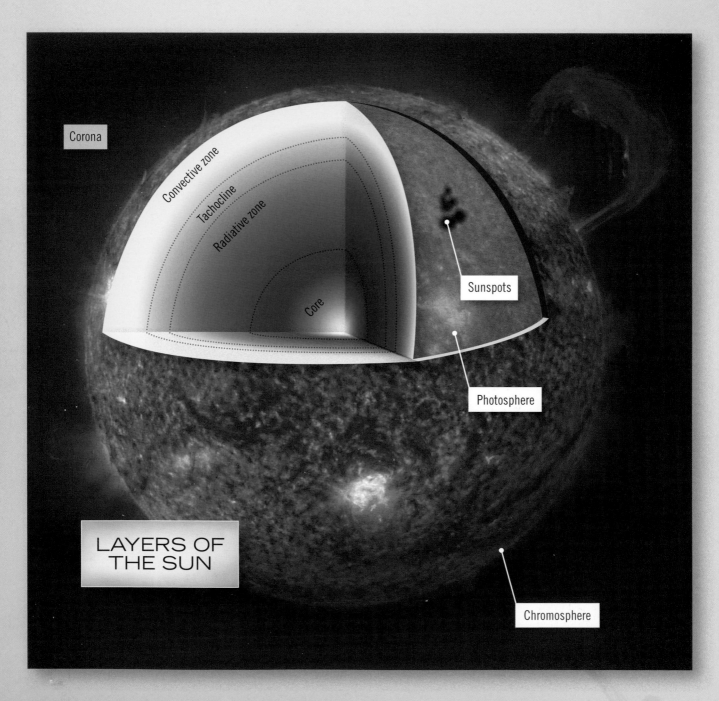

Corona

Convective zone

Tachocline

Radiative zone

Core

Sunspots

Photosphere

LAYERS OF
THE SUN

Chromosphere

The core is also the most compressed part of our Sun, with an average density of 150 grams per cubic centimeter (86.7 ounces per cubic inch), about 150 times denser than water. This is also far denser than most solid materials on Earth, about eight times the density of gold, for instance. Still, you could not stand on the Sun's core. The immense gravitational pressure exerted from the center of the Sun creates so much heat that matter cannot take solid shape.

THE RADIATIVE ZONE

The radiative zone stretches from the outer core to about 70% of the radius of the Sun. This area is known as the radiative zone because it is where energy is transported through radiation from the core outward. Superheated ions of hydrogen and helium in the radiative zone emit photons, little packets of energy, which travel a short distance until they are absorbed by a neighboring ion. This cycle of emission and reabsorption continues as the photons move in random paths from the core outward, a process that can take some photons millions of years. As plasma travels, it cools from around 15 million degrees Celsius (more than 27 million degrees Fahrenheit) near the core to about 1.5 million degrees Celsius (about 2.7 million degrees Fahrenheit) at the outermost boundary of the zone. The density of the radiative zone drops as well, from about 20 grams per cubic centimeter (11.6 ounces per cubic inch) near the core to only 0.2 grams per cubic centimeter (.12 ounce per cubic inch) at the outer edge.

THE CONVECTIVE ZONE

The convective zone extends from the outer edge of the radiative zone to the photosphere, the visible surface of the Sun. As the name implies, the convective zone is the area of the Sun's interior where the plasma is not dense enough to allow energy to be transferred through radiation. As density (and temperature) decreases, ions start holding on to absorbed photons, so nature turns to convection instead. In this zone, thermal columns carry hot material out to the surface, where it cools and then plunges downward again to absorb more energy from the base of the zone.

In the tachocline, the relatively thin boundary between the radiative and convective zones, the mechanism of energy transfer transitions from radiation to convection. Scientists believe that the Sun's magnetic field is generated by the change in the flow rate of solar material that occurs in this transition area.

THREE WAYS TO TRANSFER ENERGY

In simple terms, energy can be transferred in three ways: conduction, radiation, and convection. Conduction transfers energy between adjacent atoms that literally vibrate against one another, creating a kind of domino effect. Conduction works best in solid materials, where atoms are closely packed together.

Radiation transfers energy by passing it in tiny packets between atoms. As an atom becomes excited, it releases a packet of energy that travels until it hits a nearby atom. The neighboring atom absorbs the energy, becomes excited, and releases it again in a game of atomic hot potato.

Convection transfers energy through the flow of mass in liquid-like currents. As atoms slip and slide past one another, energy moves from "hotter" places to "cooler" places in an attempt to reach equilibrium. Since convection does not require atoms to be neatly packed against one another, it is usually how energy is transferred through less dense materials.

THE PHOTOSPHERE

Technically, the photosphere is the point at which visible light (photons) from any star can finally escape into space. Since our sight depends on detecting photons that have traveled from the Sun, the photosphere is as far "into" the Sun as we can see with the naked eye (if it were even possible to look directly at the Sun). Because photons cannot escape until they reach the photosphere, even if we could see deeper into the Sun, everything below the photosphere would appear pitch-black.

This image taken by the solar optical telescope on NASA's Hinode satellite shows a greatly magnified section of the photosphere. The light areas are granules, where hot plasma is rising up from below; the dark areas show where cooler plasma is sinking back down.

Because it is the innermost layer that can be seen, the photosphere is considered the visible "surface" of the Sun. It is about 300 kilometers (about 186 miles) thick, though its thickness varies considerably in places since it is continuously changing. The photosphere is composed of millions of granules, the very tops of convection cells where plasma at an average temperature over 6,000 degrees Celsius (10,832 degrees Fahrenheit) has risen through the convection zone and peeked out just far enough to spit visible light into the universe. After the granules release this energy, the plasma cools and falls back inward again. Each granule is about 1,000 kilometers (over 620 miles) in diameter, and each lasts about 20 minutes. The continuous coming and going of these granules gives the photosphere a constantly shifting turbulence, like the bubbling surface of a boiling stew.

THE CHROMOSPHERE

Just above the photosphere lies the chromosphere, the lowest layer of what is considered the Sun's atmosphere. If the photosphere is the surface of a boiling stew, the chromosphere is where the stew spits and splatters. It is where solar prominences (filaments of superheated plasma) anchored in the photosphere whip about.

The chromosphere is nearly 2,000 kilometers (about 1,243 miles) thick. Its density, however, is far less than the photosphere and just a tad thicker than Earth's atmosphere. In fact, under normal conditions the chromosphere is completely invisible. During a total solar eclipse, however, the chromosphere is revealed as a thin red or pink ring around the Sun's surface. Temperatures in the chromosphere vary widely, growing from nearly 6,000 degrees Celsius (10,832 degrees Fahrenheit) at its inner boundary to almost 20,000 degrees Celsius (more than 36,000 degrees Fahrenheit) at its outer edge.

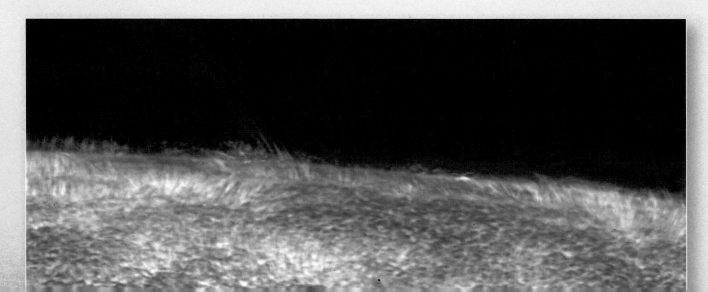

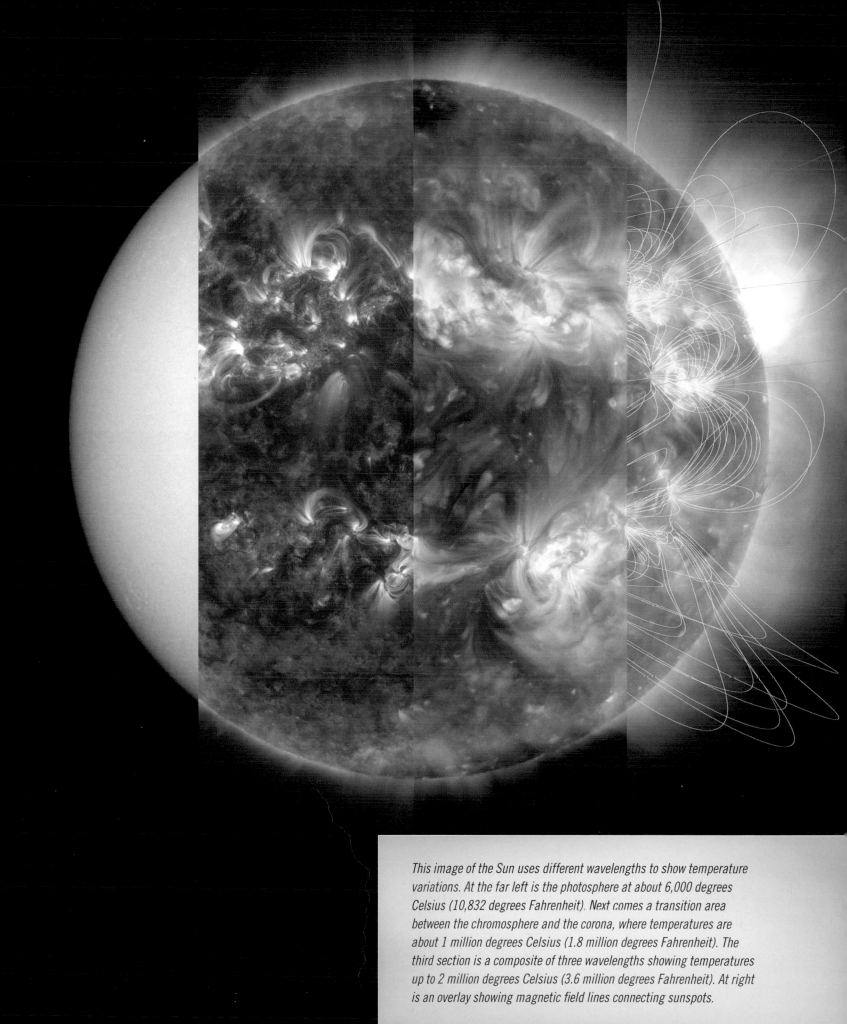

This image of the Sun uses different wavelengths to show temperature variations. At the far left is the photosphere at about 6,000 degrees Celsius (10,832 degrees Fahrenheit). Next comes a transition area between the chromosphere and the corona, where temperatures are about 1 million degrees Celsius (1.8 million degrees Fahrenheit). The third section is a composite of three wavelengths showing temperatures up to 2 million degrees Celsius (3.6 million degrees Fahrenheit). At right is an overlay showing magnetic field lines connecting sunspots.

HOW HOT IS OUR SUN?

Because of the interaction of gravitational and magnetic forces within the Sun, its temperature varies in some surprising ways. In fact, because its temperature is so variable, the more accurate question is how hot is the hottest part of the Sun? The Sun's core is about 15.7 million degrees Celsius (over 28 million degrees Fahrenheit). It is the hottest part of the Sun, by a lot. Solar material cools as it moves outward through the consecutive layers of the Sun. Curiously, however, temperatures soar again as solar material expands out through the lower layers of the Sun's atmosphere. At the chromosphere's inner boundary, the average temperature is only about 6,000 degrees Celsius (10,832 degrees Fahrenheit). But temperatures in the corona can reach as high as 10 million degrees Celsius (more than 18 million degrees Fahrenheit).

For years the fact that temperatures rose within the Sun's atmosphere baffled scientists. Recently, however, NASA's High-Resolution Coronal Imager (a small telescopic camera the agency launched on a 10-minute flight just above the Earth's atmosphere) revealed bundles of magnetically charged plasma twisting through the corona in massive braids. NASA scientists speculate that the bending and twisting of these magnetic braids interacts with magnetic field lines along the surface of the Sun. The field lines are constantly trying to straighten the magnetic braids in a process known as magnetic reconnection, which can generate enormous amounts of energy. Scientists think the extra energy released during magnetic reconnection may be heating the Sun's atmosphere and could account for the tremendous temperature difference between the Sun's surface and its corona.

NASA's High-Resolution Coronal Imager captured this image of our Sun's corona, revealing braided plasma (upper left), on January 21, 2013.

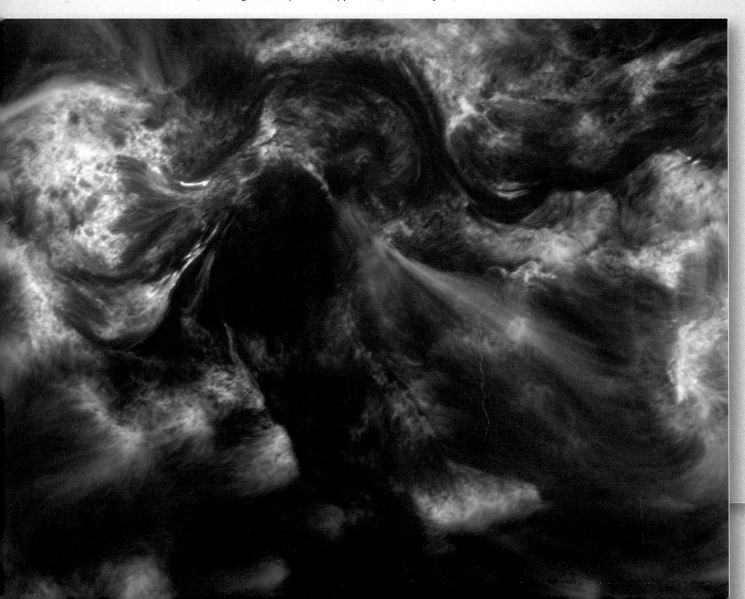

THE CORONA

We can imagine the corona as the outer layer of the Sun's atmosphere, extending millions of kilometers from the Sun's surface. It is the wispy halo visible around the Sun during a solar eclipse. It is not always evenly distributed around the Sun. Most of the actual material that makes up the corona is concentrated close to the chromosphere in the form of loops and arches of magnetically charged plasma. During periods of low solar activity, coronal material tends to accumulate near the Sun's equator, leaving the poles exposed. During active periods, the corona is distributed more evenly, covering the Sun from its equator to its poles.

The corona can reach temperatures of nearly 10 million degrees Celsius (over 18 million degrees Fahrenheit), far hotter than the surface of the Sun. This fact was discovered in the mid-twentieth century when scientists found evidence of ionized iron (which can only be formed at superhigh temperatures) in the spectral signature of light emanating from the corona.

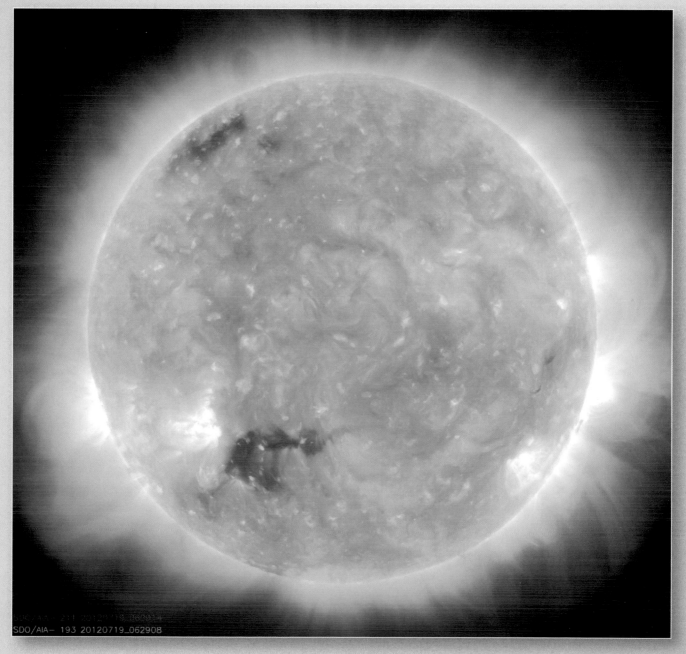

SDO/AIA– 193 20120719_062908

The corona is easily seen in this image of our active Sun from NASA's Solar Dynamics Observatory.

DISCOVERING THE SOLAR WIND

A British astronomer named Arthur Eddington first suggested the existence of what we now call the solar wind (though he never referred to it as a "wind" per se). Although Eddington presented his theory to the Royal Astronomical Institute in 1911, few astronomers took notice. In the mid-1950s, Eugene Parker, an American astrophysicist, observed that the tails of comets always pointed away from the Sun. He speculated that this was due to a wind of charged particles blowing outward from the Sun. Parker's hypothesis was met with tremendous opposition, so much, in fact, that his first paper suggesting the phenomenon was rejected for the prestigious Astrophysical Journal. *Finally, as scientists accumulated more and more evidence supporting the existence of a solar wind, Parker's theories gained widespread acceptance, and he was elected to the National Academy of Sciences in 1967.*

The fast solar wind originates from coronal holes, like the one visible as a large dark area near the top center of the Sun in this image. Coronal holes are associated with open magnetic field lines and are often found at the Sun's poles.

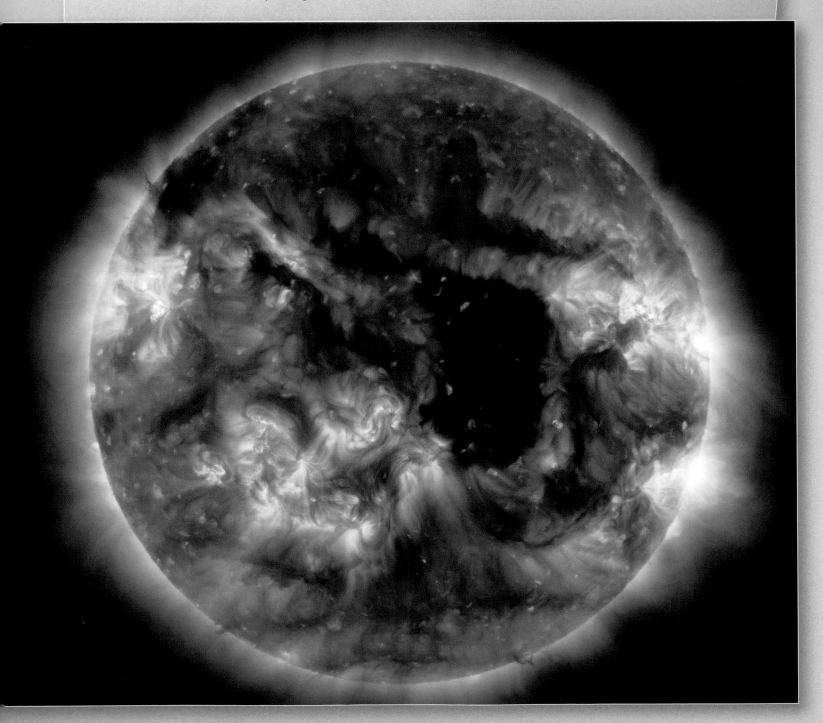

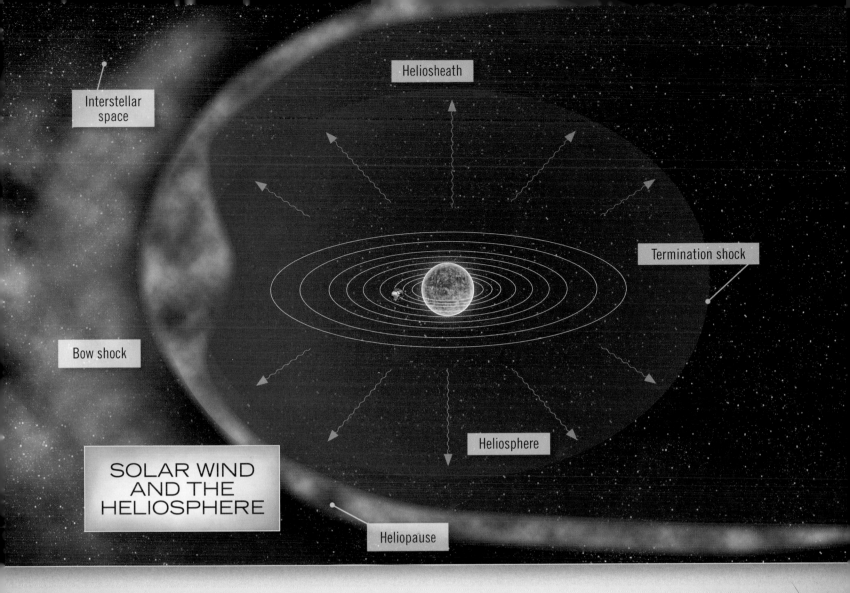

Interstellar space

Heliosheath

Termination shock

Bow shock

Heliosphere

SOLAR WIND AND THE HELIOSPHERE

Heliopause

THE SOLAR WIND

Every hour, 4 to 6 billion tons of charged particles in the Sun absorb so much energy that they escape its gravitational pull and flow outward from the corona in a low-density emission known as the solar wind. This continuous flow of particles is made up mostly of negatively charged electrons, some positively charged protons, and an unknown number of neutrally charged (or not charged at all) neutrinos. Much like wind on Earth, the solar wind constantly varies in speed, temperature, and density. This variability is largely due to the mix and total number of the particles within it.

Most scientists refer to the solar wind, but actually there are two. The fast solar wind originates from magnetically active areas near the Sun's poles and has an average velocity of about 750 kilometers (466 miles) per second. It is made of particles that resemble the elemental composition of plasma from the Sun's photosphere and has an average temperature of about 800,000 degrees Celsius (over 1.4 million degrees Fahrenheit).

The slow solar wind originates from magnetically active regions on the Sun's surface near its equator and has

an average velocity of about 400 kilometers (almost 250 miles) per second. It is nearly twice as hot and twice as dense as the fast solar wind. The slow solar wind is made of particles that resemble the elemental composition of plasma from the Sun's corona (with a higher concentration of particles from heavier elements) and has an average temperature of 1.5 million degrees Celsius (over 2.7 million degrees Fahrenheit).

The solar wind carries the Sun's charged particles and magnetic field outward in all directions, forming the heliosphere. The wind slows down abruptly as it meets interstellar space, a boundary known as the termination shock. Eventually, the Sun's magnetic field presses against the interstellar medium and bends back, forming the heliosheath, a sort of tear-shaped bubble around the solar system with the trailing edge pointed in the direction opposite the Sun's path as it moves around the galaxy. Scientists now think the compression of the Sun's magnetic field at the leading edge (the bow shock) forms immense bubbles that are swept back along an area known as the heliopause.

THE SUN'S ROTATION

The Italian astronomer and inventor Galileo Galilei was the first to realize that the movement of spots on the surface of the Sun was caused by its rotation. In 1613, he made a series of sketches of sunspots on three consecutive days and realized that they had steadily moved from left to right across the face of the Sun. In the 1860s, the British astronomer Richard Carrington went even further and used the movement of sunspots to calculate the rate of the Sun's rotation. Surprisingly, Carrington's observations showed that spots near the equator of the Sun revolved every 25 days, while spots about halfway toward the poles revolved every 28 days. What accounts for the difference is a phenomenon known as differential rotation. In fact, differential rotation accounts not only for the difference in the speed of sunspots, but for their very existence. It turns out that differential rotation may be the single most important cause of the Sun's powerful and tempestuous magnetic field.

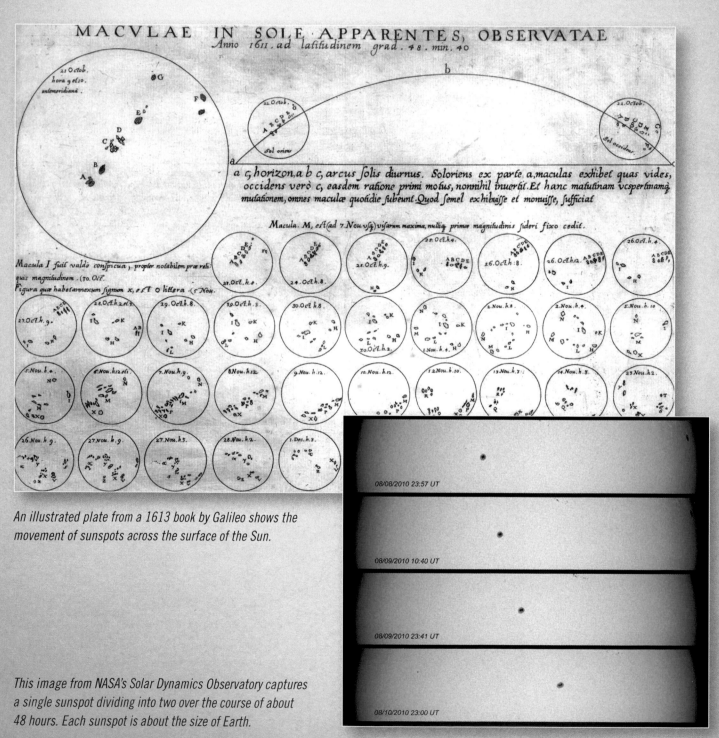

An illustrated plate from a 1613 book by Galileo shows the movement of sunspots across the surface of the Sun.

This image from NASA's Solar Dynamics Observatory captures a single sunspot dividing into two over the course of about 48 hours. Each sunspot is about the size of Earth.

08/08/2010 23:57 UT

08/09/2010 10:40 UT

08/09/2010 23:41 UT

08/10/2010 23:00 UT

DIFFERENTIAL ROTATION

Imagine spinning a perfectly spherical snow globe and observing what happens to the bits of white "snow" within it. If you watch closely, you'll notice a relationship between the location of a piece of snow inside the globe and how fast it is rotating. The snow near the equator rotates briskly, while snow near the poles hardly rotates at all. What you are witnessing is differential rotation, the phenomenon of different parts of a spinning object rotating at different rates depending on their latitudinal position.

Understanding what causes the Sun's differential rotation is a little more of a challenge. With a solid sphere, it is easy to understand that matter at the equator rotates faster than matter near the poles. The circumference of the equator is greater than the circumference of a line of latitude near a pole. Since the sphere is solid, all the matter has to complete the circumference at the same time. So something at the equator must travel faster than something near the poles.

But what happens in a fluid sphere, where matter is not connected in a way that requires it to all to rotate together? If a fluid sphere allows different parts to rotate at different rates, what determines the rotation rate for any given part? The answer involves a basic law of mechanics called the conservation of angular momentum.

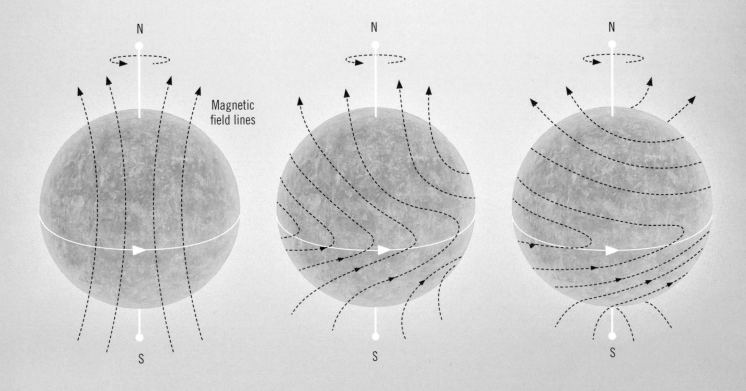

ANGULAR MOMENTUM

When force is applied to spin a fluid object, every bit of matter that makes up the object is given angular momentum, the product of how fast matter in a particular location is rotating around a central axis and how much matter (mass) is in that location. The critical point to understand is that angular momentum (a force) is related to two factors: rotation rate and mass.

It may be easier to understand the relationship between these factors when we write the definition as an equation:

Angular Momentum = Rotation Rate x Mass.

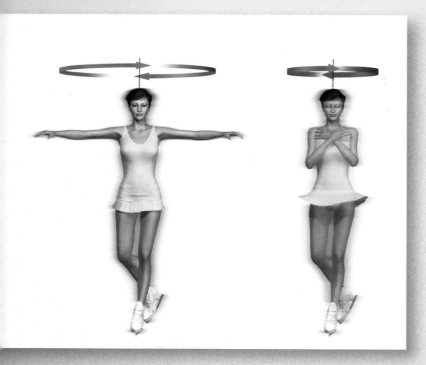

Conservation of angular momentum simply means that, for our purposes, we assume that the total amount of force that is spinning the object remains constant. No matter how we move matter within the object, the total amount of angular momentum can't change. If, therefore, we were to move some mass from the outer edge of a spinning object to somewhere close to the center, the entire object should spin faster. Since less force is needed to move mass at the outer edge faster than mass inward toward the axis (to make the full rotation in the same time), more force is available to move all the matter faster. Since the total force has to stay within the object, it can't be used for anything else.

In solid objects, it is easy to see that this is precisely what happens. Take, for example, a spinning figure skater. As she moves her arms closer to her body (the axis), she spins faster. But let's examine why. We know that matter farther away from a spinning axis (the equator) has to travel faster to cover a larger circumference in the same time as matter closer to the axis (the poles). To do so, it must use a larger amount of force.

But how much more? The answer depends on how much matter has to make the faster trip. The more matter has to move, the more force is required. Imagine our skater stretching her arms out again. We expect that her spinning will slow. Now imagine we (deftly) handed her a set of dumbbells as she continued her spin. We know that she will spin even slower (probably much slower). Thanks to the conservation of angular momentum, we now know why.

When we handed the skater dumbbells, we increased the amount of mass that had to be moved over a longer distance. That requires force. Conservation of angular momentum tells us that the amount of force we can apply to spinning the entire object is limited. If more force is consumed moving mass over a longer distance, then less is available to apply to the rate of rotation. Without more force, the rate of rotation must slow.

In solid objects, even if you can move matter around (like stretching out our skater's arms), generally you can't change the period of rotation. Because all the matter is tightly connected, it all must complete the entire rotation in the same time. In fact, the constant period of rotation is why, in a solid object, if you move matter to a location that has a larger rotational circumference, the spin of the object must slow. You don't have the option of allowing some matter to make the rotation at one time while the rest completes it at another.

Fluids, however, are not so limiting. Matter within fluid spinning objects can have different rotation periods. As a result, as matter within sloshes around, the object's total angular momentum can be distributed in all sorts of crazy ways. This uneven distribution of force ends up being the cause of much of the Sun's magnetism (as well as some of its strangest features).

DIFFERENTIAL ROTATION IN THE SUN'S INTERIOR

For a long time scientists assumed that the differential rotation observed on the Sun's surface extended into the inner layers of the Sun. However, as instruments for observing the Sun have grown in sophistication, they have allowed scientists to study wave oscillations inside the Sun. This process, known as helioseismology, revealed a surprising fact about the Sun's internal structure. From the tachocline outward, solar plasma behaves much like a fluid, exhibiting differential rotation. From the tachocline inward (the radiative zone through the core), solar material behaves more like a solid, rotating at a constant rate.

You might expect that the fluid dynamics of the Sun would result in a slower rotation period for matter at the equator than matter at the poles. After all, if less force is required to rotate matter near the poles (and there is less matter there anyway), shouldn't the force conserved increase the rotation rate of matter at the poles? Instead, just the opposite is the case. Matter at the equator rotates faster.

While fluids can conserve force with differential rotation, the angular momentum of fluid spinning objects is more easily affected by the internal movement of mass. In the Sun, convection currents are constantly moving matter outward from the core. Scientists believe the movement of this mass redistributes angular momentum toward the outer edges of the Sun, though not evenly. As matter travels outward, it also moves through layers of the Sun that are rotating at different rates. The combination of these vectors generates its own force, called the Coriolis effect, which may account for why matter at the Sun's equator appears to rotate faster than matter near its poles.

THE SUN'S WOBBLE

A little secret you are unlikely to find in any simple science textbook is that the Sun is not, in fact, at the center of our solar system. The truth is that the planets do not, strictly speaking, revolve around the Sun, either. Rather, each planet revolves around a center of mass, a different point in space where its gravitational pull is equal to the Sun's gravitational pull. For the smaller planets, this point is somewhere within the Sun's surface, but outside its center. For the larger planets, like Jupiter (at more than 300 times the mass of the Earth), the balancing point lies outside the Sun altogether. Jupiter literally revolves around a point near the Sun, not within the Sun. The combination of all of these different rotation points causes the Sun to wobble in crazy (but, thankfully, tiny) ways.

CHASING OUR SUN

As Earth rotates around the Sun every 365 (or so) days, it also follows the Sun as it speeds around the galaxy at nearly 232 kilometers (144 miles) per second. Even at this breakneck pace, it takes 240 million years for the Sun (and the rest of the solar system) to circle the Milky Way.

Just as Earth follows the Sun around the Milky Way, the Sun follows the Milky Way on its travels through intergalactic space. No one can say for sure at what speed the Sun (and the rest of us) are dragged along with the galaxy because speed is relative and, at the scale of intergalactic travel, there is really no common reference point. As much as scientists are learning about our universe, they still cannot identify its center or its edges—or, for that matter, whether it has either.

Our Sun

THE SUN'S MAGNETISM

Understanding the Sun's magnetism is the key to understanding almost all solar activity. A magnetic field is created by the flow of electrically charged particles, and inside our Sun there are a lot of electrically charged particles. Moreover, they are flowing in all sorts of directions, responding to competing forces pushing and pulling them. Scientists are still working out the exact cause (and effects) of the Sun's magnetic field, but new data from NASA's Solar and Heliospheric Observatory (SOHO) mission is giving scientists a much clearer idea of how the process works.

Scientists now think that the Sun's magnetic field originates in the tachocline, the boundary between the Sun's radiative and convective zones. Plasma from the tachocline outward acts much like a fluid, exhibiting differential rotation as the Sun spins. Plasma from the tachocline inward acts more like a solid, rotating at a constant rate. The change in the physical characteristics of the plasma at this boundary makes the tachocline much like the painted line on a highway separating the passing lane from the traveling lane. The difference in velocity between the charged particles in one lane and the charged particles in the other creates a shearing force and generates electromagnetism.

THE HISTORY OF MAGNETISM

For thousands of years, humans have known about magnetism. The Greek philosopher and mathematician Aristotle recorded discussions about magnetism by Thales of Miletus, a philosopher from Asia Minor who lived from 624 to 546 BCE. The Indian surgeon Sushruta, who practiced during the fifth century BCE, is known to have used magnets for medical purposes. In a Chinese book dating from the fourth century BCE, the anonymous author refers to the magnetic properties of lodestone (which we know today as magnetite).

Still, little was known about what caused magnetism. In 1600, William Gilbert, a British physicist and natural philosopher, published a book entitled De Magnete, Magneticisque Corporibus, et de Magno Magnete Tellure (On the Magnet and Magnetic Bodies and on the Great Earth Magnet) *in which he speculated that the Earth itself was a magnet. Of course Gilbert was correct, but he didn't know why.*

It took an accident by a Danish physicist, Hans Christian Ørsted, before anyone finally understood that there was a relationship between electricity and magnetism. During a lecture in 1820, Ørsted noticed that a compass needle reacted every time he switched a battery on and off. This chance discovery caused him to experiment with electrical and magnetic charges, and three months later, he published the first paper to posit that an electric current produces a magnetic field as it flows through a wire.

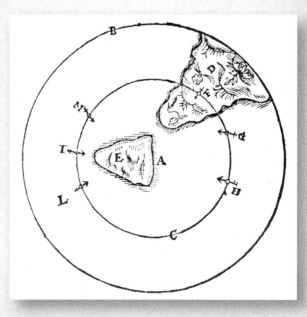

A diagram illustrating the behavior of a magnet at different points around Earth's North Pole from William Gilbert's seventeenth-century book on magnetism

Magnetic flux (one measure of the quantity of magnetism) is the product of forces acting upon moving particles, trying to move them in different directions called vectors. As charged particles rotate around the Sun's axis, they are also moving outward, traveling from the core to the surface. As they flow subject to competing vectors (particularly at the tachocline), the particles generate greater magnetic flux.

NASA scientists speculate that the magnetic field generated at the tachocline forms "tubes of force" that travel in field lines from the tachocline outward. Since different parts of the Sun are rotating at different rates, the magnetic field lines get wrapped and twisted in a chaotic mess, like a giant bowl of (very) hot spaghetti. This constant winding and bending subjects the charged particles to forces trying to push and pull them through more vectors, generating even stronger magnetism.

In 2017, NASA will launch the Solar Orbiter (SolO), a spacecraft designed to take precise measurements of the Sun's magnetic properties. Scientists hope the data from SolO will unlock the mystery of our Sun's magnetic field and provide concrete answers about electromagnetic forces deep within the star.

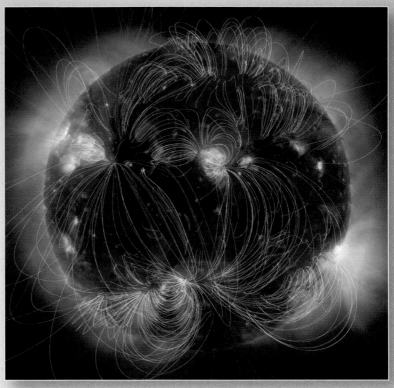

A map of the magnetic field lines emanating from the Sun is superimposed on an extreme ultraviolet image from NASA's Solar Dynamics Observatory.

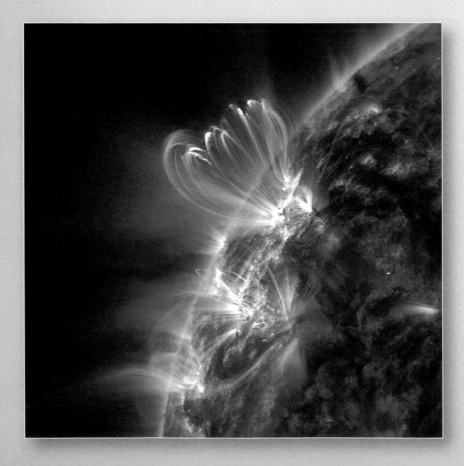

Following a solar flare, graceful magnetic loops of superheated plasma, each the size of several Earths, arc above the Sun's surface.

On July 19, 2012, NASA's Solar Dynamics Observatory captured a dazzling display known as coronal rain. After an eruption, hot plasma in the corona cooled and condensed, outlining the magnetic field lines as it slowly fell back to the Sun's surface.

In late September 2012, a series of magnetically active regions appeared to dance the conga across the surface of the Sun.

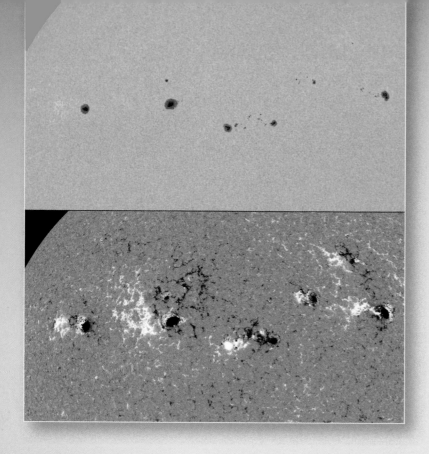

SUNSPOTS

The combination of the Sun's rotation and convection currents within the Sun twists subsurface magnetic field lines, increasing the intensity of the magnetic force following these lines. Sometimes the force grows so intense that it punches through the photosphere, momentarily blocking the normal flow of plasma through convection cells. Where this happens, surface temperatures suddenly cool relative to surrounding areas. The cooler areas appear as dark spots, known as sunspots, on the surface of the Sun.

Typically, sunspots occur in pairs with opposite magnetic polarities. The "leading" sunspot travels in the direction of the Sun's rotation ahead of the "trailing" sunspot. If you picture a giant horseshoe magnet just below the surface of the Sun, the sunspot pairs would correspond to the poles of the magnet. One sunspot exhibits a positive polarity. The other exhibits a negative polarity.

Above: *Over three days in mid-March 2013, the Sun doubled its number of sunspot groups. The bottom image, taken at the same time, shows the Sun's magnetic field, with the lightest and darkest areas indicating the strongest magnetic forces.*

Right: *A group of sunspots can be seen as bright areas near the horizon. The temperature of the glowing gases flowing around the sunspots is over 1 million degrees Celsius (2.8 million degrees Fahrenheit).*

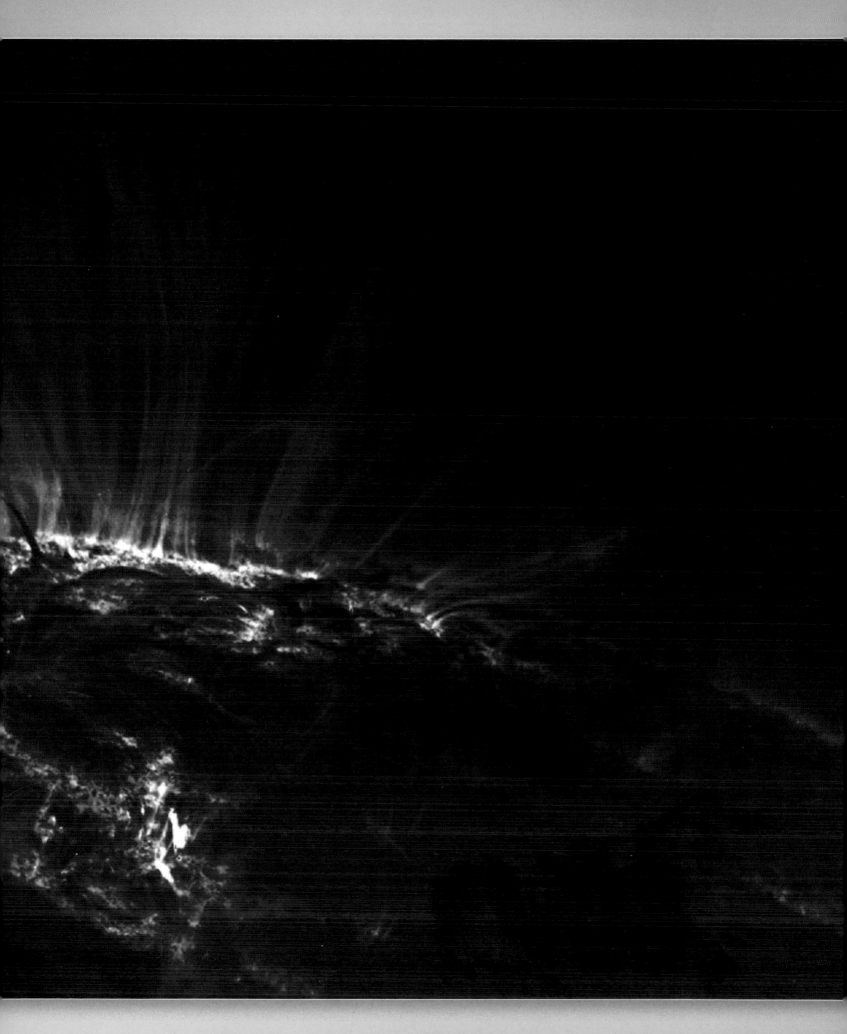

The configuration of polarities for sunspots appearing in the same hemisphere at any particular instant is always the same. That is, if the leading sunspot has the positive polarity, then the leading sunspot in all pairs that appear in the same hemisphere will be positive. Likewise, the configuration of sunspot polarity is always the opposite of the pole in its hemisphere. Leading sunspots in the Sun's northern hemisphere, for example, will have a south-oriented polarity, and the leading sunspots in its southern hemisphere will have a north-oriented polarity.

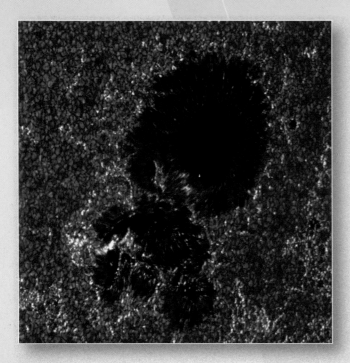

On December 14, 2006, the Hinode Solar Optical Telescope observed active sunspots during a solar flare. The granules of the photosphere are easily seen.

Generally, sunspots range in size between 1,500 kilometers (932 miles) and 50,000 kilometers (31,068 miles) in diameter. But most sunspots change size and shape during their lifetimes. Some sunspots can grow as large as 80,000 kilometers (about 50,000 miles) in diameter, large enough to swallow the planet Saturn with room to spare. The largest sunspots are visible without the aid of a telescope, though you would not be able to see them with the naked eye (and should never try).

Sunspots generally last anywhere from 1 to 100 days. When groups of sunspots appear, they typically stick around for about 50 days. Thanks to the tedious recording of sunspots by astronomers over several centuries, we have meticulous records of sunspot activity. These records show that sunspots follow a clear 22-year cycle. The number of sunspots grows over 11 years (known as solar maximum). Then, the numbers diminish over 11 years (known as solar minimum).

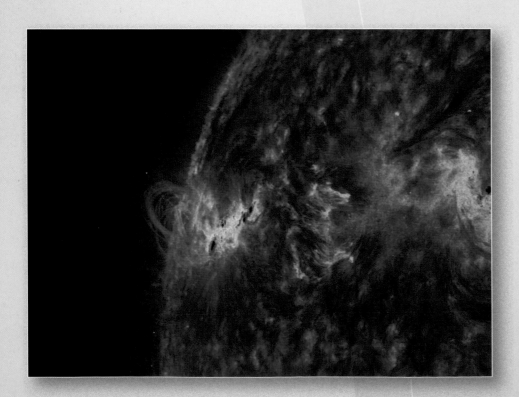

The sunspots that were the source of an X1.2-class solar flare on May 15, 2013, are clearly visible in this image from NASA's Solar Dynamics Observatory.

Ever since scientists have been able to accurately record solar radiation, measurements indicate that the number of sunspots correlate to the intensity of solar radiation. However, sunspots themselves have little impact on the Sun's radiation output. The number of sunspots also correlates to the amount of other solar activity. When more sunspots appear, so do more solar flares and more coronal mass ejections. Because of this correlation, sunspot activity is often used to predict when solar storms are likely to disrupt conditions on Earth or in near-Earth space.

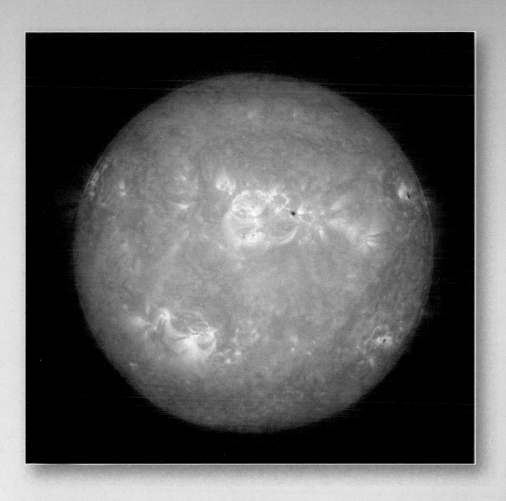

Above, right: *In this layered image of the Sun, you can peer down through the atmosphere to see the correlation of sunspots to brighter active regions above the surface.*

Right: *A pair of active regions of the Sun appear in different wavelengths, showing plasma at cooler (left) and hotter (center) temperatures as well as the sunspots (right) responsible for the activity.*

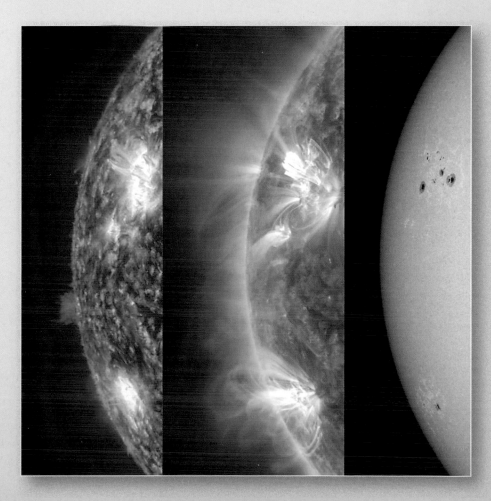

TELESCOPIC PROJECTION

The advent of the telescope made observation of sunspots much easier by using a technique called telescopic projection. If you look directly at the Sun with the naked eye, it will not take long for it to permanently blind you. But if you look directly at it through a telescope, it will blind you almost instantaneously. Early users of the telescope must have figured this out pretty quickly, because they learned to project the image of the Sun onto a flat surface. This technique, first invented by Benedetto Castelli, a pupil of Galileo, made even the smallest sunspots apparent to astronomers.

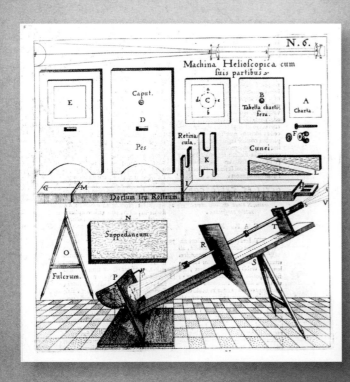

This illustration from a 1630 book by astronomer Christoph Scheiner gives instructions for building and mounting a telescope to safely observe sunspots.

WHO DISCOVERED SUNSPOTS?

Who first "discovered" sunspots is a matter of some controversy. The Chinese astronomer Gan De was the first to record their existence in his star catalogue written in 364 BCE. Not long afterward, the Greek astrologer Theophrastus mentioned sunspots in his writings on the nature of the heavens. In his book *Vita Karoli Magni* (*The Life of Charlemagne*), the Frankish scholar Einhard described a large sunspot appearing just before the emperor's death in CE 814.

Depending upon whom you ask, the first person to record sunspots observed through a telescope was the Dutch astronomer Johannes Fabricius, the Italian astronomer Galileo Galilei, the British astronomer Thomas Harriot, or the German astronomer Christoph Scheiner. In 1612, Scheiner published *Tres Epistolae Maculis Solaribus Scriptae ad Marcum Welserum* (*Three Letters on Solar Spots Written to Marc Welser*), a series of letters to a well-known patron of science that claimed Scheiner had undertaken a serious study of sunspots as early as October 1611. As a Jesuit priest, Scheiner wanted to preserve the perfection of the Sun as one of God's remarkable creations, so he incorrectly identified the spots as moons of the Sun traveling between it and the Earth.

Records indicate that Galileo may have been showing sunspots to astronomers in Rome during the spring of 1611. Sometime the following winter, Welser sent Galileo a copy of Scheiner's letters and invited Galileo to comment on them. At the time, Galileo was too ill to launch into a serious study to refute Scheiner's hypothesis.

In April 1612, however, Galileo regained his strength and embarked on a study of sunspots with the help of his student, Benedetto Castelli. Galileo concluded that the spots were either on the surface of the Sun or cloudlike structures in its atmosphere.

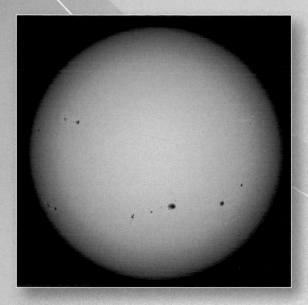

A photo of sunspots taken in summer 2012 by NASA's Solar Dynamics Observatory

THE CAMERA OBSCURA

Though Benedetto Castelli can claim credit for inventing telescopic projection, the Dutch astronomer Johannes Fabricius tracked sunspots just as accurately using a device known as the camera obscura *(Latin for "darkened room"). A camera obscura is a closed box or room with a small hole in one side. As light passes through the hole, it projects an upside-down but otherwise accurate image on the opposite inside surface of the box or room.*

Soon afterward, Galileo sent a letter to Welser reporting his findings and refuting Scheiner's claim that the spots were moons. What ensued was a flurry of letters back and forth between Scheiner and Galileo (with poor Welser in the middle). Eventually, the rivalry between the two took on a life of its own. Galileo complained about others trying to steal credit for his sunspot discoveries. Assuming the Italian was referring to him, Scheiner declared Galileo his sworn enemy.

While the two were fighting it out, apparently neither noticed that Johannes Fabricius and his father, David, had published a description of sunspots that the two had observed through a telescope David brought back from Leiden University in early 1611. But, a few months before either Fabricius or Galileo claim to have first observed sunspots, the British astronomer and mathematician Thomas Harriot recorded his observations of sunspots in December 1610 in manuscripts that were not released until after his death in 1621. Though today most books credit Fabricius as the first to observe sunspots telescopically, it was Harriot who—even according to Fabricius and Galileo's claimed dates—should be credited as the first.

No one knows who invented the device, but Aristotle is known to have used a camera obscura in the fourth century BCE to observe the crescent of the partially eclipsed Sun. The first person to ever mention a camera obscura in writing was the Chinese philosopher Mozi, who lived in the late fifth century BCE. The thirteenth-century British philosopher and Franciscan friar Roger Bacon, however, was the first to describe the use of one to safely view a solar eclipse.

This illustration, published in a 1647 book by the astronomer Johannes Hevelius, shows a telescope set up in a camera obscura in order to observe the Sun.

Over time, astronomers discovered that the smaller they made the hole, the clearer (but dimmer) the projected image appeared. Thus, by using a telescope combined with a camera obscura that had just the right-size hole, Fabricius was able to trace projections of sunspots onto paper to create remarkably accurate representations.

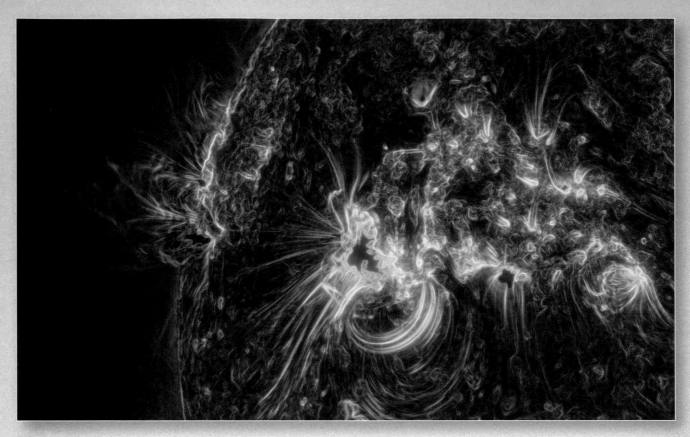

This image from NASA's Solar Dynamics Observatory was processed to enhance the visible structures. The loops represent plasma held in place by magnetic fields. Sunspots are at the center of the blue areas.

FLIPPING OUT: HELIOMAGNETIC REVERSAL

Just like the Earth, the Sun has a north pole and a south pole. But, in the Sun's case, identifying which is which may be difficult. The Sun's magnetic polarity flips about every 11 years, corresponding with a solar cycle that appears to be the result of the polarity of sunspots. The polarity configuration of sunspots is always opposite the polarity of the hemisphere in which they appear. Leading sunspots in the Sun's northern hemisphere always have a south-oriented polarity. Leading sunspots in the Sun's southern hemisphere always have a north-oriented polarity.

QUADRUPOLARITY?

When the Sun went through heliomagnetic reversal in early 2012, some scientists found unexpected magnetic changes. Rather than change from positive to negative polarity, the south pole appeared to be maintaining its positive polarity. This bizarre phenomenon led to speculation that, rather than flip, the unusual flow of magnetic material would form four poles, with two new poles near the Sun's equator. The north and south poles were expected to be positively charged, while the equatorial poles were expected to be negatively charged. Although there is no evidence of this quadrupolarity yet, one team of Japanese researchers believe that a quadrupolar pattern appeared during a heliomagentic reversal in the late seventeenth century and corresponded to the coldest part of a miniature ice age known as the Maunder Minimum.

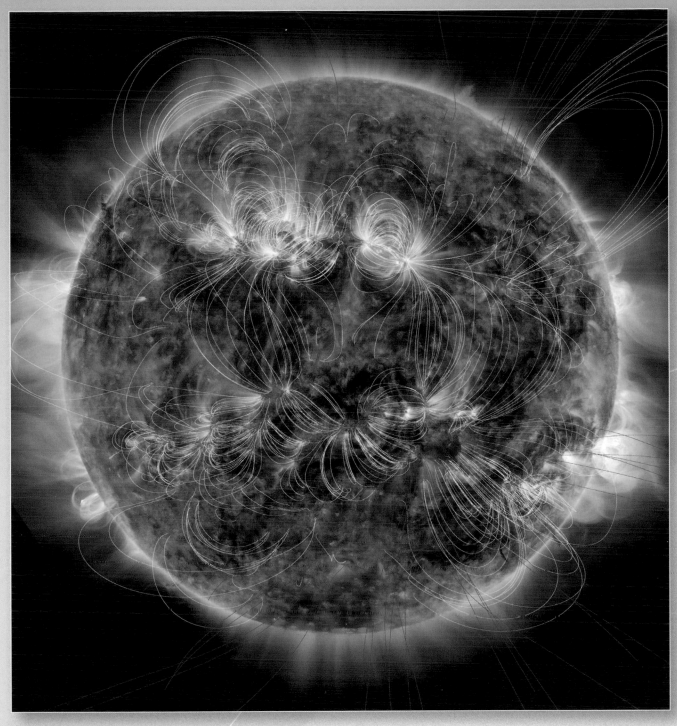

This image of the Sun was taken in the 171 wavelength by the Advanced Imaging Assembly (AIA) in NASA's Solar Dynamics Observatory on April 4, 2012. Magnetic field lines have been added to map the connections between active areas.

As more sunspots appear and expand, the polarity of the leading sunspots in a hemisphere will weaken the polarity of the pole in that hemisphere. For example, south-oriented sunspots will weaken the polarity of the Sun's north pole. Eventually, the polarity of the leading sunspots in a hemisphere will establish dominance, causing the Sun's poles to flip in a phenomenon known as heliomagnetic reversal.

The solar wind extends the Sun's magnetic field far out into the solar system. As a result, heliomagnetic reversal affects material far away from the Sun. But due to the distance involved and the speed of charged particles that make up the wind, it can take years for the effects of the Sun's "flip out" to reach all the way to the outer edges of the solar system.

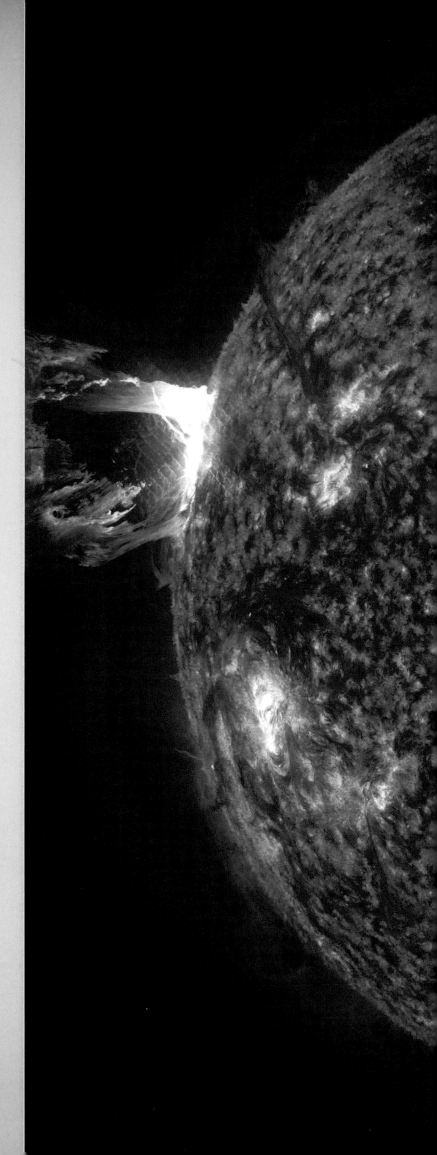

SOLAR FLARES

A solar flare is a brilliant flash on the Sun's surface associated with a sudden release of an enormous burst of electromagnetic radiation, energy equivalent to as much as 160 billion megatons of TNT. Although they release radiation across all wavelengths of the electromagnetic spectrum, most solar flares are spread over frequencies outside the range of visible light. As a result, even while many solar flares do not create a flash that can be seen with the naked eye, they can still release huge amounts of energy in the form of ultraviolet radiation, X-rays, or gamma rays.

Although scientists understand the basics of solar flares, there continues to be much debate about the details. Scientists agree that solar flares occur when electromagnetic fields within the Sun grow in intensity and attempt to escape the Sun's gravity. Most of the time, the electromagnetic forces bubble up through the Sun's surface as great loops of magnetic energy. Plasma (made of superheated, charged particles) follows the same paths as magnetic energy, forming loops of solar material known as solar prominences that jet out from the Sun's surface.

Scientists speculate that sometimes one end of a magnetic field loop will separate from the surface of the Sun and whip about like an untied shoelace. Eventually, the separated end reconnects with the surface, closing the magnetic loop. However, occasionally the loose end will reconnect by entering the surface where a neighboring loop connects. If the strength of the magnetic force in the neighboring loop is weaker when the loose end reconnects, it will release the difference as excess energy.

On April 16, 2012, a spectacular medium-size solar flare and prominence eruption occurred simultaneously.

In early January 2013, charged particles spun along magnetic loops emanating from a pair of significantly active regions along the Sun's surface.

The excess energy released during magnetic reconnection heats up charged particles in the Sun's corona, accelerating them to near the speed of light. These particles emit the energy as massive flashes of radiation, from radio waves to gamma rays. Consequently, flares are classified as A, B, C, M, or X, according to the amount of energy they release. Each class is further characterized along a scale ranging from 1 to 9. Thus, an X1 flare belongs to the most energetic class but is four times less powerful than an X5 flare.

X-class flares can cause radiation storms in the Earth's atmosphere and impact radio transmissions and other wireless communications. They also pose significant health risks to any astronauts orbiting the Earth, since their spacecraft are not as shielded by the Earth's magnetosphere.

In 2013, NASA's Solar Dynamics Observatory (SDO) managed to snap the first photos of the magnetic reconnection process in action. These images may help scientists confirm the cause of solar flares and devise better ways of predicting when they will occur.

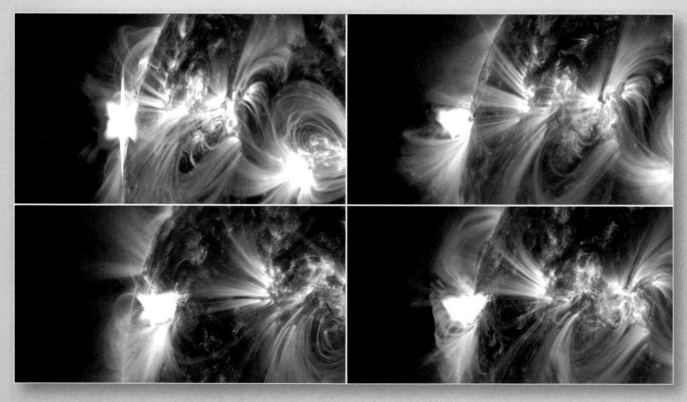

The Sun emitted the first four X-class flares of 2013 on May 12–14. Clockwise from top left, the flares were classified as X1.7, X2.8, X3.2, and X1.2.

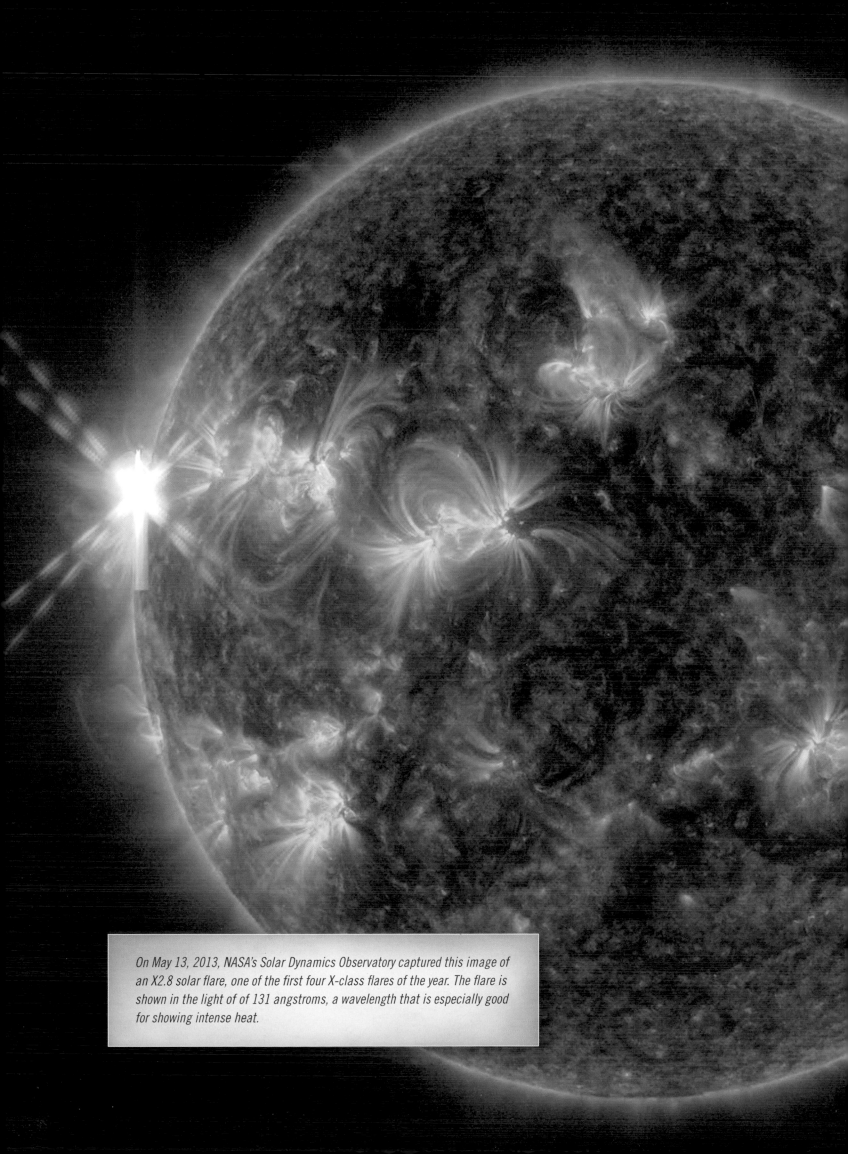

On May 13, 2013, NASA's Solar Dynamics Observatory captured this image of an X2.8 solar flare, one of the first four X-class flares of the year. The flare is shown in the light of of 131 angstroms, a wavelength that is especially good for showing intense heat.

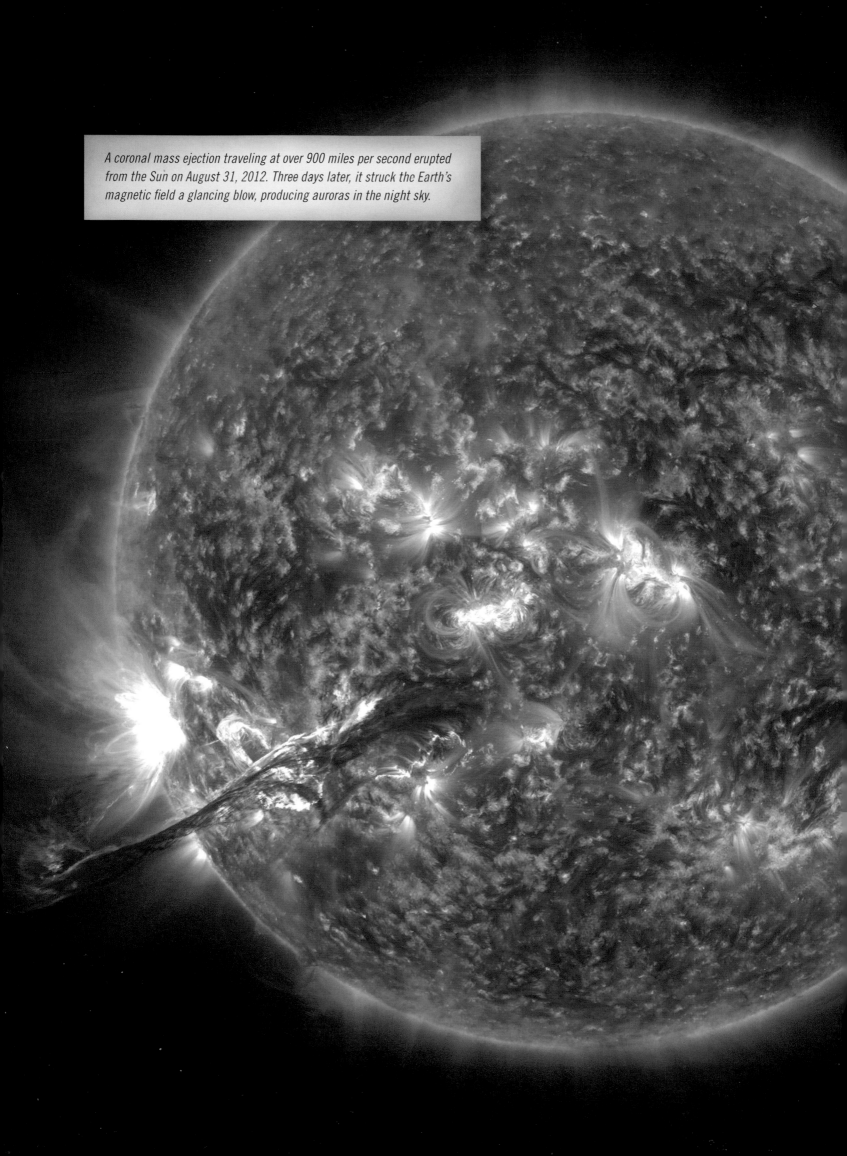

A coronal mass ejection traveling at over 900 miles per second erupted from the Sun on August 31, 2012. Three days later, it struck the Earth's magnetic field a glancing blow, producing auroras in the night sky.

CORONAL MASS EJECTIONS

Coronal mass ejections (CMEs) are massive bursts of energy and solar material flung from the Sun into space. They differ from solar flares mostly in scale (CMEs are stronger) and form (CMEs always involve solar material, while flares may not). It is easy to confuse the two phenomena. Even some astronomers conflate them on occasion. Both are thought to be triggered by magnetic reconnection on the Sun's surface. Strong flares are often accompanied by CMEs, but each can occur in the absence of the other.

When currents of magnetically charged plasma try to erupt from the Sun's surface, they are usually thwarted by the Sun's sheer mass. Gravity forces magnetic field lines to bend back toward the Sun's surface, dragging streams of plasma with them. Sometimes, however, as a magnetic field line is bent back into the Sun, it will suddenly snap like a dry twig, releasing tremendous amounts of energy. This violent reaction hurls a mass of charged particles into space and, sometimes, directly toward the Earth. A large CME can contain a billion tons of electrically charged matter in a long trail of solar spit traveling close to the speed of light, although most CMEs travel much slower.

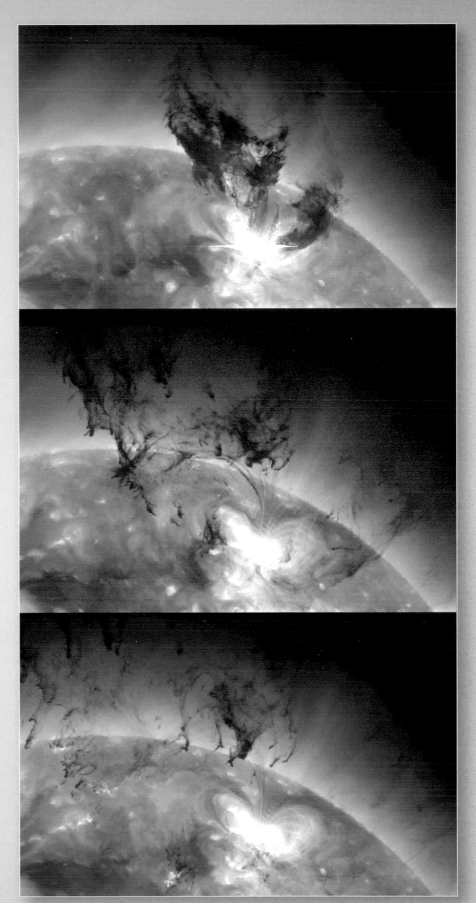

On June 7, 2011, the Sun unleashed a medium-size flare and a huge coronal mass ejection. These three stills from NASA's Solar Dynamics Observatory were taken over a period of just 30 minutes.

On September 1, 1859, Richard Carrington, a British astronomer, was watching sunspots when he observed a sudden bright point of light on the Sun's surface grow brighter and then slowly dim. The next day, the Earth experienced a massive electromagnetic storm. Bright auroras appeared as far south as the Caribbean. Carrington suspected that there might be some relationship between the storm and the solar activity he had observed the day before and reported his theories to the Royal Astronomical Society. The episode would come to be known as the Carrington Event, the strongest recorded geomagnetic storm ever sparked by a CME.

The first photograph of a CME was taken on December 14, 1971, by the Orbiting Solar Observatory (OSO-7), the seventh in a series of nine satellites NASA launched between 1962 and 1975 to study the Sun. The image was digitized to seven bits, compressed, and transmitted to the Naval Research Laboratory. The full, uncompressed image would have taken almost 44 minutes to transmit.

NASA recorded the fastest CME on April 14, 2012. The Solar Terrestrial Relations Observatory (STEREO), two nearly identical satellites launched in 2006 to provide three-dimensional images of the Sun, clocked a CME traveling between 2,900 and 3,200 kilometers (1,800 to 2,000 miles) per hour just as it erupted. Though the CME eventually slowed, it took less than 17 hours to blow past the Earth.

SUN QUAKES

Magnetically charged material belched from the Sun's surface often rains back down into the chromosphere. Sometimes these spouts of plasma rebound with such force they create sun quakes, seismic waves that ripple across the surface of the Sun just like quakes on Earth. Astronomers predicted sun quakes as early as 1972. But it wasn't until a team of NASA scientists analyzed 1996 data from the SOHO mission that anyone observed one. The picture—which wasn't verified until 1998—was of an 11.3-magnitude quake, about 40,000 times more powerful than the famous 1906 quake that leveled San Francisco. Still, few pictures of sun quakes exist, although that may change as NASA expands its observations of the Sun over the next few years.

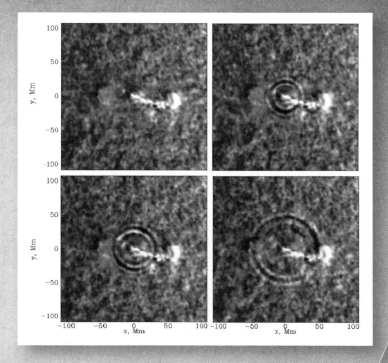

NASA's Solar and Heliospheric Observatory captured the first image of a sun quake in 1996.

In this image of a coronal mass ejection from NASA's Solar and Heliospheric Observatory, an occulting disk was used to block out the Sun's glare. Superimposed on the disk is an image of the Sun, scaled to its approximate size.

The biggest flow of electrically charged particles in our solar system is, of course, the solar wind. As you can imagine, the movement of all that charge creates a very large magnetic field. In fact, the interplanetary magnetic field it creates extends all the way through our solar system until it reaches interstellar space.

The combination of the Sun's differential rotation and the solar wind's dynamic pressure affects the shape of this interplanetary magnetic field, twisting it into a spiral pattern that undulates like the skirt of a ballroom dancer as she whirls about. In fact, the field rotates along with the Sun, making a full revolution every 25 days on average. As it does so, the folds of this magnetically charged skirt interact with charged particles in the Earth's magnetosphere and create beautiful auroras in the skies near the North and South Poles.

INTERPLANETARY MAGNETIC FIELD

The dividing line between the inward magnetic field direction of one hemisphere of the Sun and the outward direction of the other forms a neutral current sheet. This sheet undulates as the solar wind carries the magnetic field outward in a spiral pattern created by the Sun's rotation.

Opposite page: An illustration of a particle cloud blasted out from the Sun is superimposed over a photo of an aurora taken by an astronaut on the International Space Station.

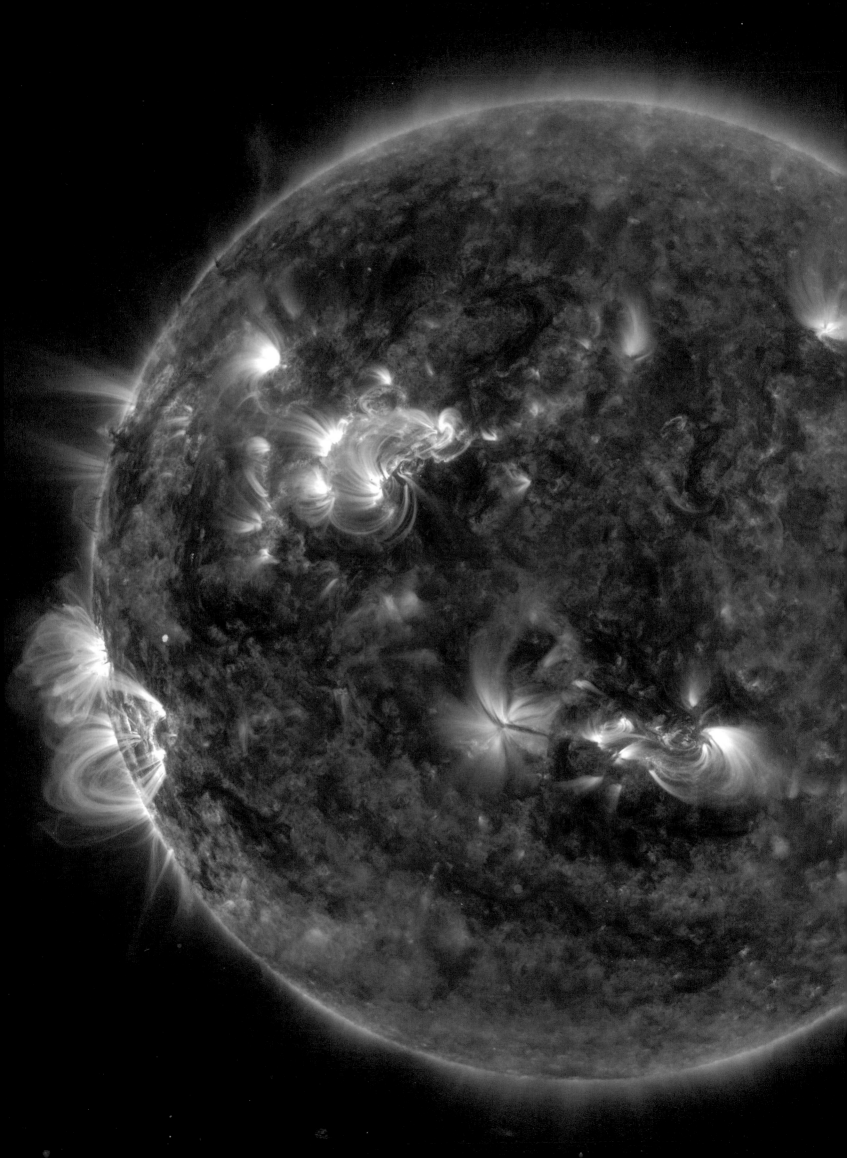

3

THE IMPORTANCE OF OUR SUN

Our Sun is one giant ball of energy. That's true. But it is also so much more than that. Before science revealed the secrets of how our Sun functions, humans knew its value and praised its worth. Nothing else in the sky was as brilliant, and nothing else brought light and warmth to their otherwise cold and dark world.

Our early ancestors didn't know how, but they knew the Sun was intimately connected to the fertility of the Earth. Before they knew that the Earth revolved around the Sun, they knew that it determined the seasons, bringing the scorching summer heat, the autumn harvest, the winter snow, and the rebirth of spring. Technology may have expanded our understanding of the Sun, but its allure has not diminished. Our Sun remains as important to humanity today as when the first humans gazed upon its splendor and wondered . . .

HOW FAR AWAY IS THE SUN?

You might think that the answer to this question would be pretty straightforward. But for thousands of years, astronomers and mathematicians have attempted to calculate the distance between the Earth and the Sun, arriving at wildly different estimates. Over 2,000 years ago, the Greek astronomer and mathematician Aristarchus of Samos calculated the distance to be between 18 and 30 times the distance between the Earth and the Moon. As we now know, his method was not only inaccurate but also impractical. Since its formation nearly 4.5 billion years ago, the Moon has slowly but steadily been moving farther and farther away from the Earth, while the distance between the Sun and the Earth has remained relatively constant. So even if Aristarchus's method was accurate, it would require constantly measuring the changing distance between the Earth and the Moon and updating that ratio to reflect the distance to the Sun.

Almost a hundred years after Aristarchus, Archimedes, another Greek astronomer, claimed that Aristarchus actually estimated the distance between the Earth and the Sun using only geometry. According to Archimedes, Aristarchus calculated that the distance was equal to about 10,000 times the radius of the Earth. We know now that the distance is more than 23,000 times the radius of the Earth. If Archimedes was reporting accurately on the earlier astronomer, then Aristarchus's estimation still wildly missed the mark.

The astronomer Claudius Ptolemaeus (Ptolemy)

In the second century CE, Claudius Ptolemaeus (Ptolemy), an Egyptian astronomer and mathematician of Greek descent, tried to calculate the distance between the Earth and the Sun using a new (for his day) form of mathematics: trigonometry. But he made a critical mistake. He tried to estimate the apparent sizes of the Sun and the Moon and concluded that, as seen from Earth, the diameter of the Sun must be approximately equal to the diameter of the Moon. Using this ratio, and complex calculations of the shadows that the Earth cast on the Moon during lunar eclipses, Ptolemy computed the distance between the Earth and the Sun to be about 1,210 times the radius of the Earth. Of course, Ptolemy's estimate was way off, too. Nevertheless, for over 1,500 years, no one really challenged it.

Finally, in the seventeenth century, Western civilization underwent a revolution in scientific thinking. The German astronomer and mathematician Johannes Kepler questioned Ptolemy's calculations and realized they must be too low—much, much too low. Telescopes had been invented in Kepler's time, which allowed for far more accurate measurements of angles between the Earth and the stars than Aristarchus and Ptolemy could make using the naked eye. Telescopes allowed astronomers to calculate distances between celestial objects using a technique called parallax.

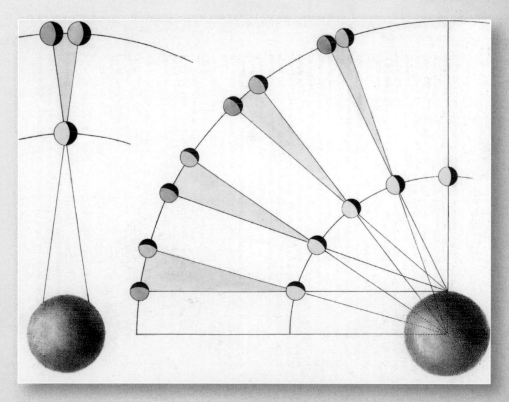

A mid-nineteenth-century diagram illustrating planetary parallax

Imagine you are speeding down the highway in your car. You glance down at the needle on your speedometer and see that you are traveling at 75 miles an hour, dangerously exceeding the posted speed limit of 65 miles an hour. "Oh, dear!" you exclaim to the friend comfortably snuggled into the passenger seat. "I really need to slow down." Your friend, glancing over at the speedometer, replies, "Nonsense! You're only going 65." The reason you worry and your friend does not is because of parallax, the different position of the speedometer needle when viewed from two different lines of sight. From your position directly in front of the speedometer, you know that you are speeding along at 75. But from his vantage point, the speedometer needle appears to be right at the 65-mile-an-hour mark.

It turns out that astronomers can use parallax to measure the relative positions of celestial bodies and, thereby, the distance between them. Christiaan Huygens, a Dutch astronomer and contemporary of Kepler's, used telescopic measurements of parallax to estimate that the distance between the Earth and the Sun was equivalent to about 24,000 times the radius of the Earth, remarkably close to modern measurements. Giovanni Cassini, an Italian, came even closer. Using trigonometry and studying Mars from two different points on Earth, Cassini used parallax to calculate the distance from the Earth to the Sun to within about 6%.

Modern instruments are capable of measuring distances longer than 150 million meters (about 93,200 miles) to within a few meters. But this precision presents some new challenges. For one thing, the orbit of the Earth around the Sun is an ellipse rather than a true circle. As a result, the distance between the Earth and the Sun is constantly changing. On an even more basic level, how should we determine where the Earth stops and the Sun begins? Should we measure the distance from the center of the Earth to the center of the Sun? Should we measure from the top of the highest mountain on Earth to the farthest reaches of the Sun's chromosphere?

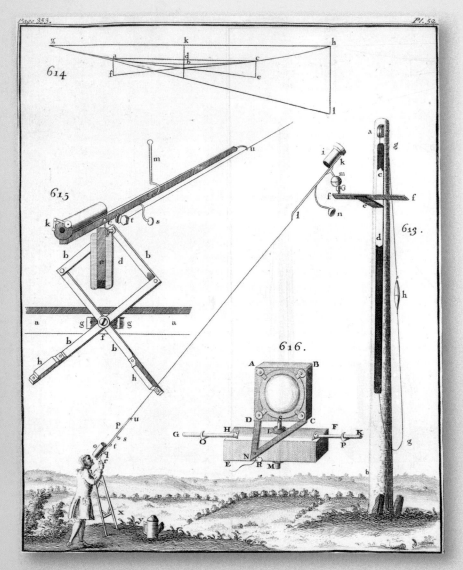

A 1738 engraving of Christiaan Huygens's 210-foot aerial telescope, which had an eyepiece connected by a taut string to a short tube mounted on a pole. The tube contained the light-gathering optics and could be maneuvered with the string.

For a long time, modern astronomers used a complex mathematical equation that involved the mass of the Sun, the length of a day on Earth, and a fixed number known as the Gaussian gravitational constant to determine the distance between the Earth and the Sun. But the Sun's mass is always changing as the fusion reactions in its core transform some of its mass into energy. To make things even more complicated, astronomers began using the astronomical unit (AU), the distance between the Earth and the Sun, to express the distance between all sorts of celestial bodies. Even a minor difference in the calculation could have huge effects when calculating the distance of far-off solar systems and distant galaxies.

If all of this sounds confusing, rest assured, you are not alone. Many scientists have desired an easier method of calculating cosmic distances. In 2012, they got their wish. The International Astronomical Union, the official authority on such things, decided to redefine the AU. Under the new definition, an AU would no longer fluctuate depending on the Sun's mass or the length of an Earth day. Instead, a group of scientists reached a consensus that the unit would be a simple constant. Today, the official distance between the Earth and the Sun (one AU) is exactly 149,597,870,700 meters (almost 93 million miles).

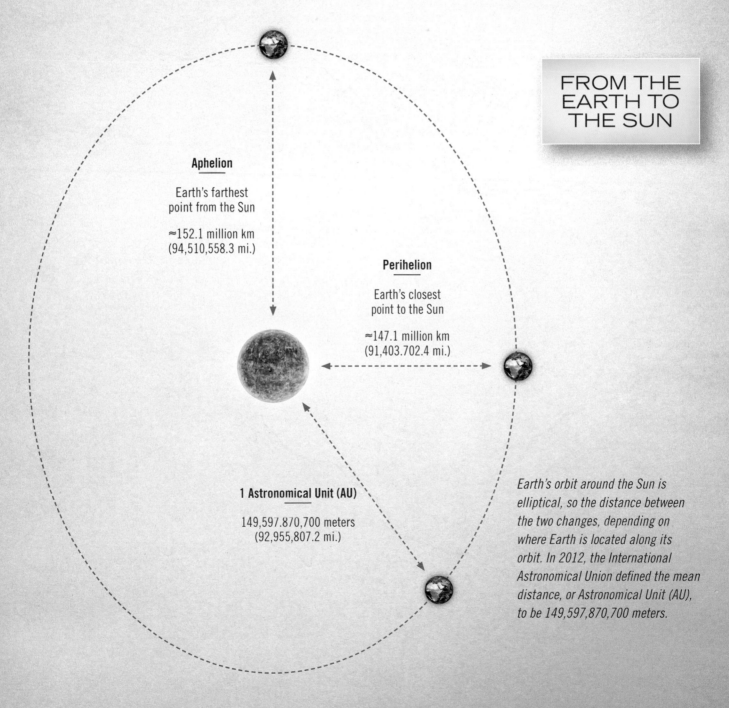

FROM THE EARTH TO THE SUN

Aphelion

Earth's farthest point from the Sun

≈152.1 million km
(94,510,558.3 mi.)

Perihelion

Earth's closest point to the Sun

≈147.1 million km
(91,403.702.4 mi.)

1 Astronomical Unit (AU)

149,597.870,700 meters
(92,955,807.2 mi.)

Earth's orbit around the Sun is elliptical, so the distance between the two changes, depending on where Earth is located along its orbit. In 2012, the International Astronomical Union defined the mean distance, or Astronomical Unit (AU), to be 149,597,870,700 meters.

THE SWEET SPOT

For human beings (in fact, for all life on Earth), there is more to this distance than simplifying astronomical measurements. Rather, 149,597,870,700 meters (almost 93 million miles) is a very special number because that distance puts the Earth comfortably within an area scientists call the circumstellar habitable zone, or CHZ.

At least as far as we know, all life-forms require an environment with water in liquid form. In fact, all life-forms on Earth are based on carbon compounds dissolved in a bath of H_2O. When scientists look for extraterrestrial life, therefore, they look for planets with water. That means planets within the CHZ. Planets too far from their star will be too cold—any water on the surface would freeze. Planets too close to their star will be too hot—any water would quickly boil away.

At 149,597,870,700 meters, Earth is in the sweet spot, the region around a star within which it is theoretically possible for a planet to maintain liquid water on its surface. This near-perfect balance is thought to be the primary reason that life took hold here.

The circumstellar habitable zone (CHZ)—or the sweet spot—is the area around a star where a planet will not be too hot or too cold to maintain liquid water on its surface. In our solar system, Earth is the only planet that orbits within the Sun's CHZ.

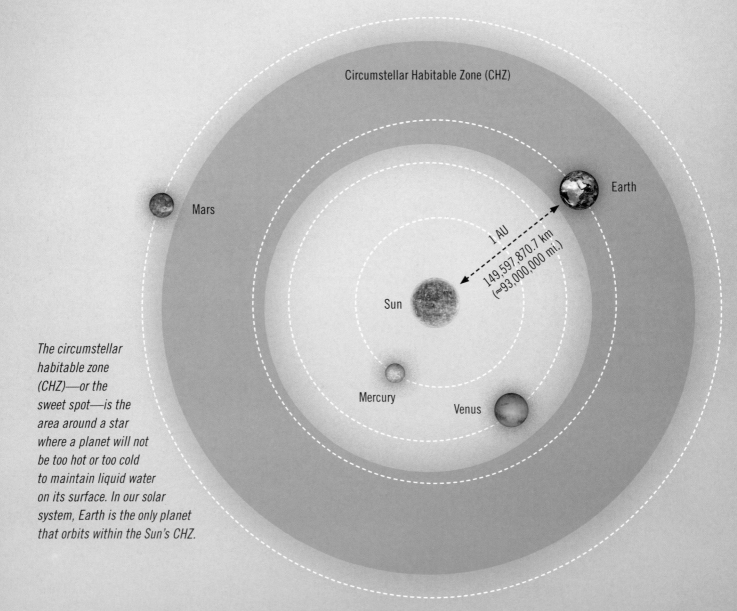

Circumstellar Habitable Zone (CHZ)

Mars

Earth

1 AU
149,597,870.7 km
(≈93,000,000 mi.)

Sun

Mercury

Venus

TITAN—OUR BEST CHANCE OF FINDING EXTRATERRESTRIAL LIFE?

While planets orbiting within circumstellar habitable zones are the most likely places in the universe to find extraterrestrial life, there is some controversy over what does and does not constitute habitability. Some scientists argue, for example, that liquid water need not exist on the surface of a planet. In fact, many scientists speculate that the best place to find extraterrestrial life within our solar system is in a place far outside the CHZ.

Titan, the largest of 60 known moons of Saturn, exists so far away from the Sun that scientists speculate any water on its surface must be in a solid state. Indeed, at –179 degrees Celsius (–290 degrees Fahrenheit), any water on Titan's surface would turn to (and stay) ice. But what about under the surface? Based on data from a flyby of Titan by NASA's Cassini spacecraft, many scientists believe that under this moon's icy surface is a vast ocean of liquid water and ammonia. Combined with data indicating Titan's atmosphere may be rich in organic compounds, these scientists believe that Titan, though far outside the official CHZ, may be the most suitable environment in our solar system to sustain some form of extraterrestrial life.

In fact, in 2009, NASA prioritized funding for the Titan Saturn System Mission (TSSM), a joint mission between NASA and the European Space Agency that would, among other things, look more closely for signs of life on Titan. The fact that so many of the world's leading astronomers and astrophysicists are convinced that life could exist there calls into question whether CHZs are the only (or even the best) places to search for life in the universe.

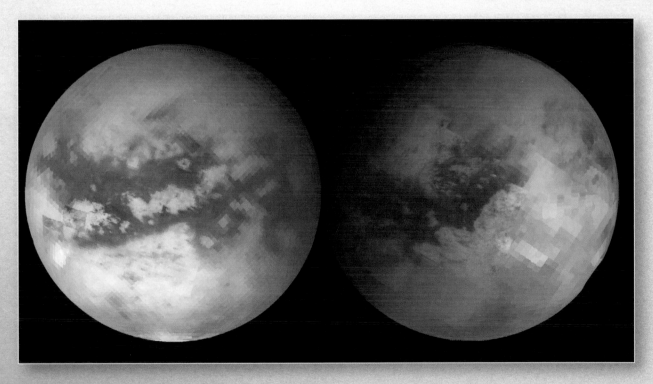

These false-color images of Titan were taken by the Cassini Orbiter in a 2005–06 flyby.

LET THERE BE LIGHT

Our Sun is the primary source of light on Earth. In fact, nearly all the light we see originated in the Sun. The gleam of the Moon and the dim glow of planets in the night sky are mere reflections of sunlight. The only other natural sources of visible light on Earth come from the few creatures that use chemical reactions to create bioluminescence, relatively rare atmospheric events like lightning, and the very, very faint glimmer of distant stars. All other visible light comes from the Sun. After taking a journey lasting anywhere from several thousand to one million years, photons from deep within the Sun's core reach its surface. From there, they take a short 8.3 minutes to reach the Earth.

Above left: *A satellite view of sunlight falling on Earth*

Above: *The Moon glows with reflected sunlight.*

An 1874 engraving of Isaac Newton experimenting with light, while his roommate, John Wickins, looks on.

But what, exactly, is light? Around 55 BCE, Titus Lucretius Carus, a Roman poet and philosopher, speculated that the light of the Sun was composed of minute particles that travel instantaneously across the universe. Over 1,500 years later, the French mathematician and philosopher René Descartes rejected this particle theory, arguing that instead, light behaved much like sound waves, traveling faster through denser mediums. Isaac Newton, a British mathematician and contemporary of Descartes, rejected the wave theory. He reasoned that, if light was a wave, it would bend around objects the way other waves do, so in theory, we would be able to see objects clearly even when they were situated behind other objects. Newton's logic was compelling enough that the majority of the scientific community embraced the particle theory of light well into the nineteenth century.

In 1918, a German theoretical physicist named Max Planck won the Nobel Prize for inventing quantum mechanics. Planck suggested that, although light was a wave, finite amounts of it gained or lost energy based on the frequency at which they vibrated. In 1926, Gilbert Lewis, an American, named these vibrating lumps of light *photons*. Essentially, quantum mechanics allowed Lucretius, Descartes, and Newton all to be correct to some degree. According to quantum mechanics, light is made up of photons that sometimes act like particles and sometimes act like waves. If you find the dual nature of photons hard to visualize, you're not the only one. Even today, physicists struggle with light's apparent schizophrenia.

In its broadest sense, sunlight is made up of an entire spectrum of electromagnetic radiation released by the Sun as photons, some of which we see, most of which we do not. In fact, we only see about 44% of the photons that reach Earth's surface. Partly, this is because sunlight is filtered through the Earth's atmosphere, which blocks some forms of solar radiation. Mostly, however, we cannot see all of the light from the Sun because many of the photons vibrate at wavelengths that are either too long (infrared) or too short (ultraviolet) for our eyes to sense.

EVOLUTION OF SIGHT

You owe the fact that you can read the words in this sentence to our Sun. By blanketing the Earth with ultraviolet radiation, some of which bounces along the surface, our very early ancestors evolved the power of sight. Whether something as complex as the eye was the product of evolution, however, has often been controversial. Even Charles Darwin, the father of the theory, noted in his book *On the Origin of Species* that the development of the power of sight through natural selection seems "absurd" at first glance. However, he went on to suggest that the complex eyes we use today evolved gradually from simple optic nerves coated with a pigment, which made them capable of sensing photons if not of perceiving light, color, or shape. All of that would develop, according to Darwin, in "numerous gradations" over millions of years.

And, indeed, that seems to be precisely how it happened. Just recently, scientists have traced the origin of sight to an ancient ancestor of the modern hydra, a tiny creature that lives in lakes, ponds, and streams in temperate and tropical climates. Over 600 million years ago, the hydra's predecessors—little more than blobs of goo floating about Earth's shallow seas—developed the ability to distinguish dark from light. Researchers from the University of California, Santa Barbara, discovered that a gene called opsin was found in the DNA of hydras but not sponges, even though sponges are the most primitive forms of all animals. The opsin gene is responsible for triggering the production of light-sensitive proteins, also called opsins, that coat the surface of the hydra and help the animals tell day from night. By tracking the opsin gene back in time, researchers were able to trace the origin of sight to early forms of a group of animals known as Cnideria, which included ancient sea anemones and jellyfish.

Scientists have traced the origin of sight to an ancient group of animals called Cnideria, which includes modern jellyfish, as shown here.

Development of the eye took place during a particularly active evolutionary period known as the Cambrian explosion. However, because the fossil record of the early Cambrian period is so poor, scientists turned to computer programs designed to model small mutations exposed to natural selection. Remarkably, these computer models found that a primitive light-sensing organ like a pigmented nerve could evolve into the modern complex eye in as little as 400,000 years.

Scientists believe that early eyes consisted of "spots" where light-sensitive opsins surrounded an area of pigment called the chromophore, which allowed primitive distinctions between levels of brightness. With these spots, the animals could detect the presence of light but not the direction it came from. The ability to determine the direction of light required the evolution of tiny cups where the eyespots were located. As light entered these cups, it would hit different opsins depending on its angle. As the cups deepened, fewer opsins would be activated by the bouncing light waves, providing a sharper and sharper indication of the direction of the light source.

Because these early see-ers lived in the ocean, they were exposed to the only two wavelengths of the electromagnetic spectrum that can penetrate water: blue and green visible light. Scientists speculate that this is the reason that eyes developed to detect only visible light—a narrow range of wavelengths within the full spectrum.

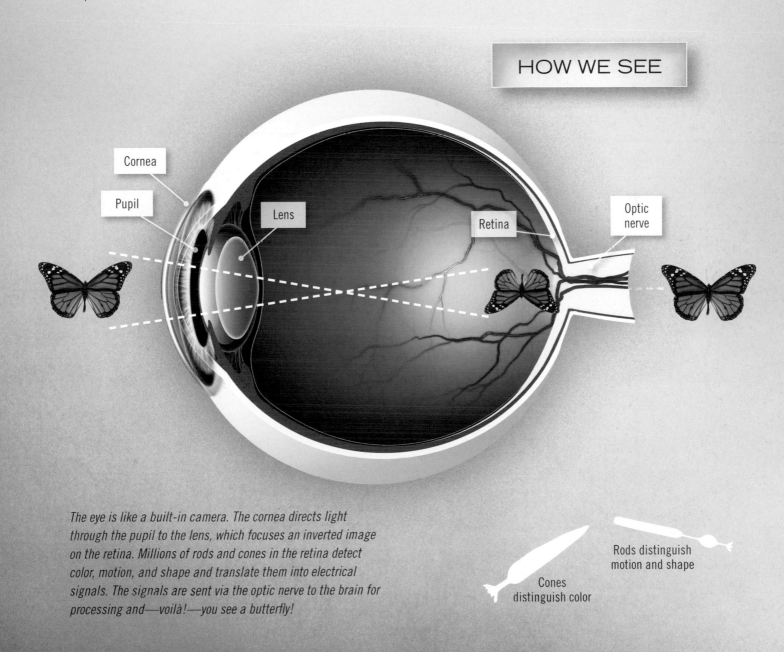

HOW WE SEE

Cornea

Pupil

Lens

Retina

Optic nerve

The eye is like a built-in camera. The cornea directs light through the pupil to the lens, which focuses an inverted image on the retina. Millions of rods and cones in the retina detect color, motion, and shape and translate them into electrical signals. The signals are sent via the optic nerve to the brain for processing and—voilà!—you see a butterfly!

Rods distinguish motion and shape

Cones distinguish color

Eventually, the advantages of primitive forms of sight (catching food, avoiding dangers, and finding mates) led to the evolution of your eyes. Though they are far more complex, the same basic principles remain. Light (photons that act like both particles and waves) enters your eye, first passing through a football-shaped transparent covering called the cornea. The cornea bends the light waves so that they focus on the pupil, the black opening in the center of your eye. The pupil is surrounded by the iris, the colored part of your eye that closes and opens as needed to let in or block out light. The light that gets past the iris passes through a lens, which both focuses and inverts the light, sending it upside down toward the retina.

It is the retina that contains the millions of photoreceptor cells covered in those same opsin proteins that sparked the evolutionary process some 600 million years ago. The opsins in your retina, however, come in two different types. Rods detect only shades of gray, but are useful for distinguishing motion and shape. Cones, which are usually concentrated toward the center of the retina, mostly detect one of three colors: red, blue, or green. As each of these different types of opsins sends a signal over your optic nerve, your brain mixes red, blue, and green into all the colors we see, much like an LED screen turns individual pixels into dramatic, multicolored images.

By developing two eyes, early animals were able to better detect distance. If you've ever taken your glasses off during a 3-D movie, you have probably noticed the overlapping, nearly identical two-dimensional images on the screen and concluded that the glasses somehow melded them both to produce a single, clear, three-dimensional image. What you may not have realized is that your eyes, not the glasses, are performing this amazing feat, and they do it all the time . . . even without glasses. Since each eye generates a slightly different image, the brain can fill in the slight differences between the two in a way that creates a single, three-dimensional image sitting sharply in space. This dimensionality helped greatly enhance depth perception—and eventually allowed us to enjoy watching our favorite superheroes leap from cinema screens.

A page illustrating binocular vision from the notebooks of the fifteenth-century Italian artist and scientist Leonardo da Vinci

THE MIRACLE (AND CURSE) OF PHOTOSYNTHESIS

One way or another, nearly all life on Earth owes the Sun a debt of gratitude. With very few exceptions, every living thing on the planet obtains its energy directly from the Sun, through photosynthesis, or indirectly, by consuming things that get their energy from the Sun. The one notable exception is a group of very simple, single-celled microorganisms called Archaea, which usually can be found living in hot springs and volcanic mud holes or in hydrothermal vents on the ocean floor. As inhospitable as these places seem to us, scientists believe that the early Earth was essentially one huge, scalding-hot thermal vent. Once the planet's surface cooled enough to form a crust, volcanic gases created a thick atmosphere of mostly ammonia, methane, and water vapor (with a little bit of hydrogen, nitrogen, carbon, and oxygen thrown in). As the Earth cooled, the atmosphere condensed and fell back to the surface as a kind of toxic stew that formed the primordial ooze of the early seas. It was in these seas that life on Earth first appeared.

Archaea, single-celled microorganisms, can thrive in boiling hot, sulfurous volcanic mud holes.

While no one knows precisely what the first form of life on Earth looked like, most scientists believe it was very similar to today's Archaea: single celled, without a nucleus, and remarkably consistent in shape and size. The most notable characteristic of these first organisms is that they derived their energy completely independent of the Sun. For all these early life-forms cared (if they *could* care), the Sun could fade to a bundle of smoldering cinders. They simply didn't need it. (And most modern Archaea still don't.)

To live, all life-forms must find a way to convert the chemical energy in their environments to adenosine triphosphate (ATP), the coenzyme that cells use to transport energy *within* the cell. Transporting energy *between* cells requires an electron transport chain, something that will allow energy to cross the cell membrane. Usually these chains consist of a chemical that will trade an electron (a donor) paired with a chemical that will attract an electron (an acceptor). The whole process of converting and transporting energy is called cellular respiration. Generally, it happens in one of two ways. Aerobic respiration uses oxygen to break down and transport the energy in sugars. Anaerobic respiration uses, well, anything else (usually sulfate, nitrate, sulfur, or fumerate).

It turns out that oxygen is incredibly eager to accept electrons. It wants them. It needs them. It desires electrons so passionately that it will cut ahead of other chemicals to get them if it has to. Oxygen's constant craving for electrons makes aerobic respiration 19 times as efficient as anaerobic. In anaerobic respiration, one unit of glucose is converted to two units of ATP. In aerobic respiration, it's 38. Oxygen is so efficient at attracting electrons that we call the chemical reaction in which a substance loses electrons *oxidation*.

Oxygen has another interesting characteristic: it is the by-product of photosynthesis. Before photosynthesis, almost all life on Earth used anaerobic respiration to create energy. Sure, it wasn't particularly efficient, but free-floating oxygen was hard to come by. Earth's early atmosphere contained exactly none. What little oxygen there was on Earth was bound up in water. But all of that changed around 2.6 billion years ago (no one knows precisely when), after cyanobacteria in the primordial seas evolved (no one knows precisely how) oxygenic photosynthesis, the quirky ability to combine sunlight with carbon dioxide and water to produce ATP.

OXYGENIC PHOTOSYNTHESIS

The chemical processes involved in oxygenic photosynthesis are so complex it's a wonder life took only 400 million years to figure it out. Put simply, any pigment is capable of absorbing sunlight in the form of photons. In oxygenic photosynthesis, there are two light reactions. Oxygen is created in the first one and released into the atmosphere. When chlorophyll molecules absorb photons, they pass the energy along from molecule to molecule until it reaches a structure called a reaction center. A pair of special chlorophyll molecules called P680 absorbs the light energy, becomes unstable, and ejects one electron for every photon. The electrons are passed, two at a time, into the electron transport chain. This process of passing electrons between molecules allows energy to move across cell walls and is one way plants produce electrochemical energy.

In order to repeat the process, the electrons ejected from P680 have to be replaced. This is accomplished by splitting two water molecules into four free-floating protons, four free-floating electrons, and one molecule of oxygen:

$$2H_2O \rightarrow 4H^+ + 4e^- + O_2$$

The electrons are used by P680, the protons are later used in the production of ATP, and the oxygen is released into the atmosphere to become the air we breathe.

HARVESTING SUNLIGHT TO MAKE OXYGEN

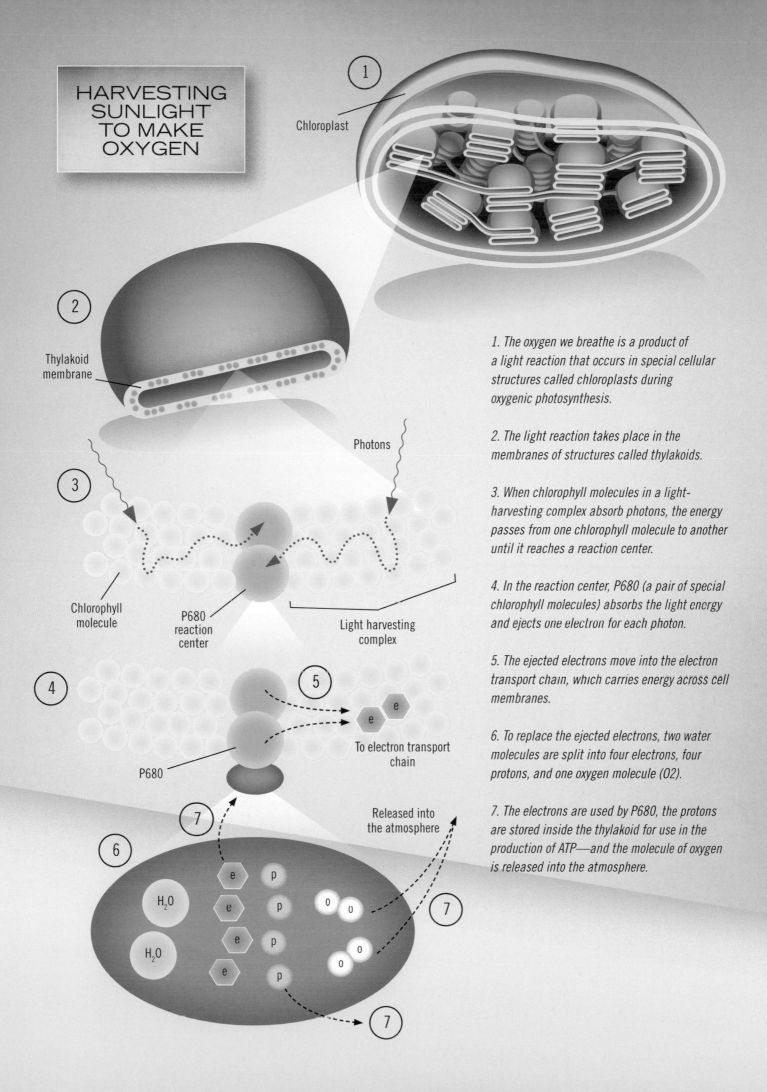

1 Chloroplast

2 Thylakoid membrane

Photons

3 Chlorophyll molecule

P680 reaction center

Light harvesting complex

4

5

e e

To electron transport chain

P680

Released into the atmosphere

7

6

H₂O

e p

e p

e p

o o

o

o

H₂O

e p

7

7

1. The oxygen we breathe is a product of a light reaction that occurs in special cellular structures called chloroplasts during oxygenic photosynthesis.

2. The light reaction takes place in the membranes of structures called thylakoids.

3. When chlorophyll molecules in a light-harvesting complex absorb photons, the energy passes from one chlorophyll molecule to another until it reaches a reaction center.

4. In the reaction center, P680 (a pair of special chlorophyll molecules) absorbs the light energy and ejects one electron for each photon.

5. The ejected electrons move into the electron transport chain, which carries energy across cell membranes.

6. To replace the ejected electrons, two water molecules are split into four electrons, four protons, and one oxygen molecule (O2).

7. The electrons are used by P680, the protons are stored inside the thylakoid for use in the production of ATP—and the molecule of oxygen is released into the atmosphere.

Cyanobacteria, similar to these found in the hot springs of Yellowstone National Park, were the first organisms capable of photosynthesis.

THE OXYGEN CATASTROPHE

Roughly seven-eighths of Earth's history comprises the Precambrian era, from the birth of the planet about 4.5 billion years ago until roughly 541 million years ago. Although very little is known about this early period, one thing is clear: anaerobic organisms like the humble Archaea dominated. They were multitudinous. They were everywhere. And for the vast majority of our planet's history, they ruled.

Then, one cyanobacterium performed a photosynthetic magic trick some 2.7 billion years ago and everything started to change. Through photosynthesis, it created energy at lightning-fast speeds (compared to all the other life-forms at the time) and released oxygen as a by-product. Oxygen began accumulating in Earth's atmosphere, slowly at first, as cyanobacteria capable of oxygenic photosynthesis gained a foothold on their genetic competition. But around 2.4 billion years ago, a critical threshold was reached and the concentration of oxygen in Earth's atmosphere exploded. Paleobiologists call this the Great Oxygenation Event (or the Oxygen Catastrophe, depending on your perspective). For anaerobic organisms, however, it was the end of an era. Oxygen is poison to them, and its rapid buildup in Earth's atmosphere was probably responsible for wiping most of them out. In fact, you could say that the miracle of photosynthesis caused the largest extinction event in Earth's history.

For aerobic organisms, photosynthesis was a godsend. It was a vastly more efficient mechanism for organisms to convert and transport energy. It allowed for larger and more complex life-forms to evolve. Some scientists believe it was the catalyst for the Cambrian explosion, the sudden appearance of dozens of diverse animal phyla around 530 million years ago. For a period of about 80 million years, the rate of evolution accelerated and helped produce the vast diversity of plants and animals that has been the hallmark of our planet ever since.

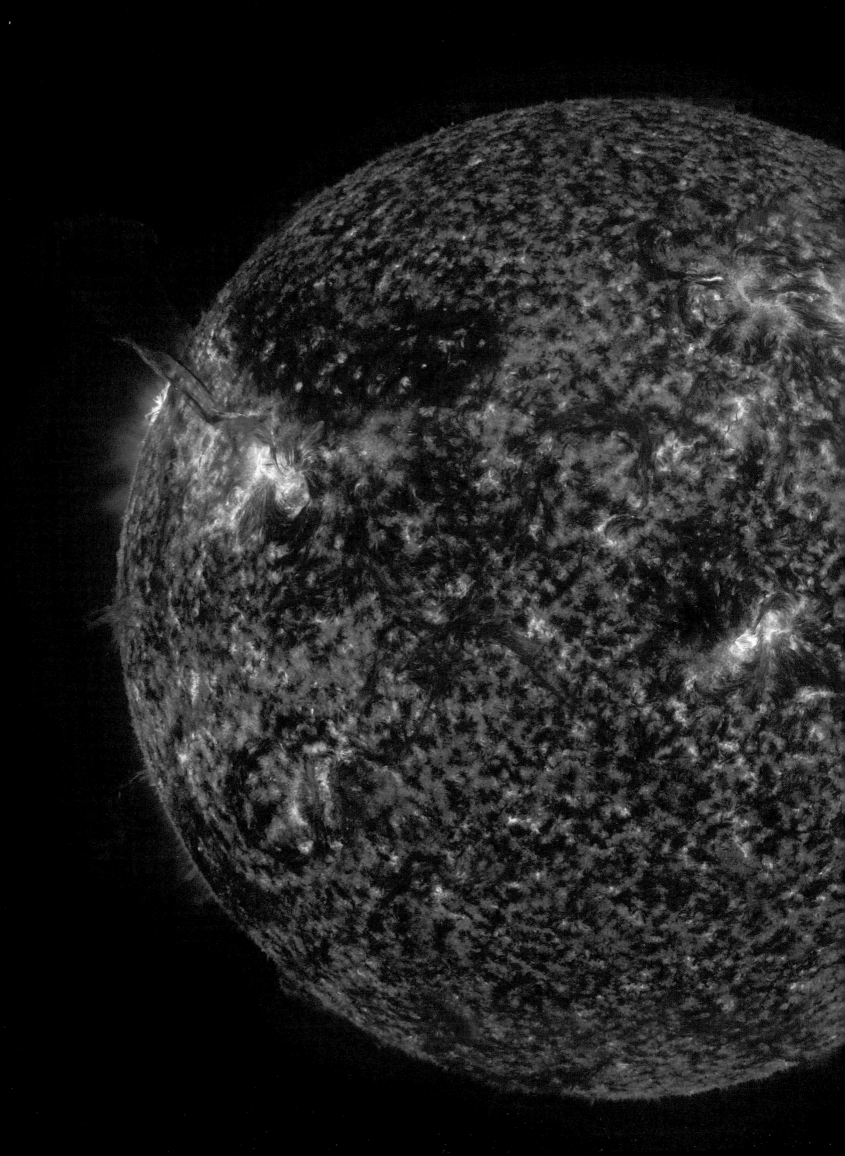

THE WORSHIP OF OUR SUN

When humans finally arrived on the scene, life's connection with the Sun took a decidedly theological turn. In fact, as far as we can tell, Sun worship is almost as old as humanity itself. Humans have venerated the Sun ever since they gazed upon the brilliance of its rising and the splendor of its setting. Nearly every culture has invented some form of Sun god. Since many civilizations sprang up with the development of agriculture, it comes as no surprise that the inhabitants of the first ones worshiped the Sun for providing daily sustenance.

Many early cultures fashioned elaborate mythologies to explain the Sun's creation and movements. Some believed the Sun traveled across the sky in a boat or a chariot. Others believed there were multiple Suns that appeared from among the leaves of a colossal cosmic tree at different points during the day.

AZTEC: TONATIUH AND HUITZILOPOCHTLI

The Aztec people believed that there was a series of Suns and had a cyclical solar calendar similar to that of the Maya. Aztec cosmology held that each Sun was a god that ruled during its own cosmic era and that their civilization was in the era ruled by Tonatiuh. Tonatiuh was the fifth Sun in the cycle and had expelled the fourth Sun from the sky in a celestial battle. According to the Aztec creation myth, the Sun would refuse to move through the sky unless it was appeased with human sacrifices. As many as 20,000 people were sacrificed each year to Tonatiuh and other gods.

Huitzilopochtli was a Sun god worshipped by the Aztecs and other Mesoamericans, including the Mexicas of Tenochtitlàn, the famed city-state located on an island in Lake Texcoco in the Valley of Mexico (the site of modern-day Mexico City). Huitzilopochtli was also the god of war and human sacrifice. He is often depicted as a blue man with a great crest of hummingbird feathers on his head. According to Aztec mythology, Huitzilopochtli's mother became pregnant with him when a ball of hummingbird feathers fell from the sky and touched her. Huitzilopochtli's siblings felt their mother's mysterious pregnancy brought dishonor upon the family. One sister, Coyolxauhqui, encouraged her siblings, who were all stars, to kill their mother. But Huitzilopochtli sprang from his mother's womb, cut off Coyolxauhqui's head, and threw it into the sky, where it became the Moon.

The Sun god Tonatiuh's face stares out from the central disk of an Aztec calendar stone.

A modern rendering of Huitzilopochtli based on images in Aztec codices

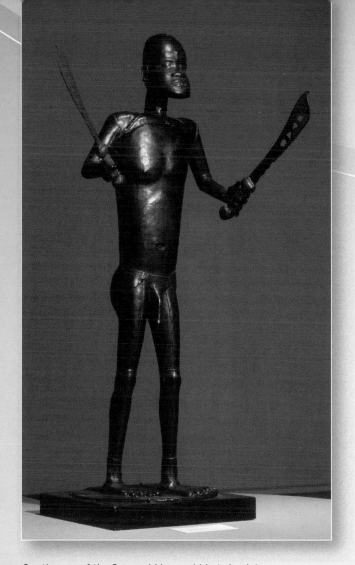

Gu, the son of the Sun god Liza and his twin sister, the Moon goddess Mawu

FON: LIZA

The Fon people of West Africa worshiped a Sun god named Liza, a boy who represented heat, work, and strength. Liza had a twin sister, the Moon goddess Mawu, who represented night and motherhood. Liza and Mawu were the children of Nana Buluku, the first mother. According to Fon mythology, the universe was created as a result of the twins' incestuous relationship, which was spurred on by a cosmic serpent named Da. After the birth of the universe, Liza is said to have fathered a son, named Gu, who was born as a divine tool in the shape of an iron sword. According to the Fon, Liza used Gu to shape the world and to teach humans how to work iron.

Twins or dualistic powers appear in many African creation myths, as does a huge snake often associated with the rainbow. The Bushongo people of the Congo region had a slightly different Sun mythology. They believed that the creator god Bumba was the sole inhabitant of a universe made of water. One day Bumba vomited up the Sun, which dried up the water, creating dry land. Bumba then vomited up the first humans to inhabit the new land.

CELTIC: LUGH

The Sun god of the Irish Celts was Lugh. He was the grandson of the god Balor, leader of the giant Fomorians, who were sometimes associated with storms, disease, and other powers of nature. According to Celtic prophesy, Balor would be killed by a grandson, so fearing for his life, Balor sent Lugh away. The young god was raised by Manannán, the god of the sea, who taught him to be an excellent warrior. When Lugh reached manhood, he joined the Tuatha Dé Danaan who were oppressed by Balor and the Fomorians.

Lugh and the Tuatha Dé Danaan waged war against the Fomorians. Balor had only one eye, but it was capable of killing whoever looked upon it. He usually kept it closed except in battle. Just as Balor was opening his eye at the battle of Mag Tured, the brave Lugh cast a stone at it. Hitting its target with immense force, the stone pushed the evil eye out the back of Balor's head, killing the god instantly and wreaking havoc on the army of giants behind him. To this day, the Celtic people venerate Lugh with an annual festival in August commemorating the beginning of harvest.

Lugh had to prove himself to Nuada, king of the Tuatha Dé Danaan, by performing several tasks, including winning at chess, as shown in this illustration from a 1905 book.

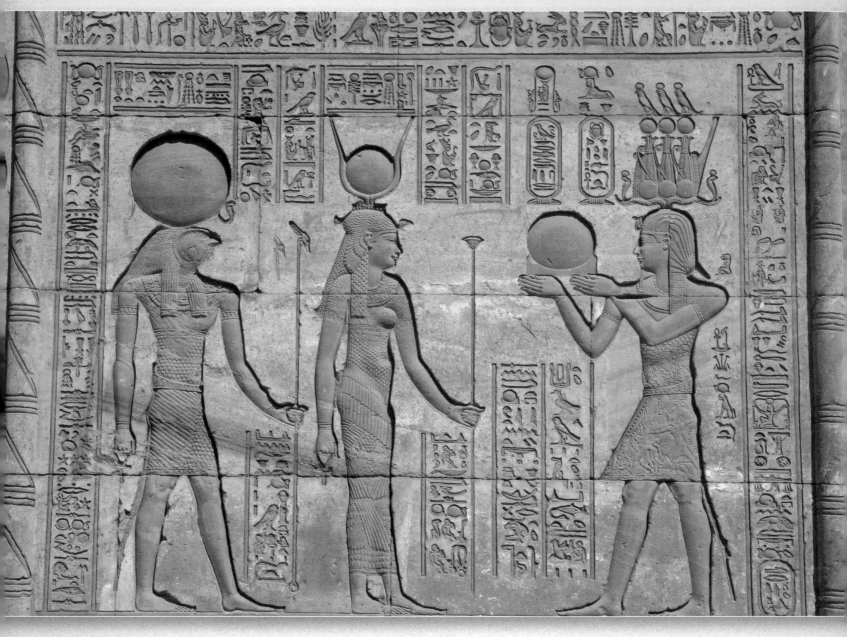

A depiction of Re-Horakhty (left) in the Temple of Hathor at Dendera, Egypt

EGYPTIAN: RE

The ancient Egyptians had many gods, but the Sun god Re (also Ra) was one of the most revered. In fact, his name may be a variant of the Egyptian word for "creator." During later Egyptian dynasties, Re was merged with the god Horus (Re-Horakhty) and was believed to rule over all parts of the world. Re was thought to have created all forms of life, bringing each into existence by calling its secret name. Among one cult of Re, worshipers believed that the god's tears created man. The Egyptian *Book of the Dead* recounts how Re cut himself and his blood transformed into the two intellectual personifications Hu (authority) and Sia (mind).

By the Fourth Dynasty, Egyptian pharaohs were declaring that they were Re, manifest on Earth. This claim caused worship of the god to soar in popularity, leading pharaohs of the Fifth Dynasty to spend the majority of their treasure building Sun temples and adorning the walls of their tombs with increasingly complex Re mythology. When Christianity became the official religion of the Roman Empire, Egyptians abandoned the worship of Re, though study of the god remained an academic interest among Egyptian priests.

CHINESE: XI HE AND THE TEN SUNS

The ancient Chinese believed that there were ten Suns, the children of the goddess Xi He and Di Jun, the lord of heaven. Xi He bathed her children in a lake located in the Valley of the Light in the East and placed them in an enormous mulberry tree, whose branches rose to the heavens. Each day, one of the Suns would journey across the sky toward the summit of Mount Yen-Tzu in the West.

One day, the Suns grew tired of this routine and decided that they would all take the journey together. But the heat of ten Suns withered the people's crops, dried up the rivers, and threatened to destroy the Earth. In desperation, Emperor Yao prayed to Di Jun, asking him to command his children to appear one at a time. But the Suns defied Di Jun, so he sent his archer, Yi, to frighten them by firing arrows from his magic bow. Instead, Yi shot all but one of the Suns. In anger, Di Jun condemned the archer to live as a mortal on Earth. The only child spared by Yi is the Sun we see today.

HINDU: SURYA

Ancient Hindus worshiped the Sun god Surya, sometimes depicted as a man with three eyes and four arms. Two of his hands were said to hold water lilies. He used a third hand to encourage worshipers to approach him, while blessing them with the fourth hand. Surya was thought to ride a chariot pulled by seven horses.

According to Hindu mythology, Surya married a goddess named Sanjna. Because she could not bear his intense light and heat, she fled into the forest and disguised herself as a mare. But Surya discovered where she had hidden. Appearing as a horse, he approached her and the two were reunited. Eventually, she bore him several children. However, Sanjna's domestic duties were made intolerable by Surya's heat. So she appealed to her father, who cut Surya's body to reduce his brightness to one-eighth of its original brilliance.

India's thirteenth-century Konârak Sun Temple represents the chariot of Surya, the Hindu Sun god, led by a team of seven horses.

INCA: INTI

The Inca of ancient Peru worshiped the benevolent Sun god Inti, believed to be the ancestor of the Inca people. Inti was married to the Earth goddess Pachamama. They had two children: a son, Manco Capac, and a daughter, Mama Ocllo. Inca mythology holds that Inti taught his children the methods of civilization and sent them to Earth with orders to teach humanity. Manco and Mama were told to build the Inca a capital city with a magic golden wedge that would fall to the ground. The Inca believed Cuzco to be this city and, even today, the Festival of Inti Raymi is celebrated there each year.

A detail of a replica of the golden disc representing Inti, the Inca Sun god

INUIT: MALINA

Ancient Inuit people believed in a Sun goddess, Malina. Her brother was the Moon god Anningan. According to Inuit mythology, the two got into a quarrel, which caused Malina to storm off in anger. Anningan chased her, whether to apologize or continue the argument remains a point of dispute. Because Anningan did not eat in his constant pursuit of Malina, the Moon waned as he grew thinner and thinner. The new Moon appeared when Anningan finally stopped to eat. Then, as Anningan resumed the chase, the Moon waxed again. Only occasionally, during a solar eclipse, was Anningan ever thought to catch up with Malina.

GREEK: HELIOS AND APOLLO

In Greek mythology, the god Helios personified the Sun. Helios was thought to be the son of Hyperion and Theia, two of the twelve Titans, the original sons and daughters of Gaia, the physical incarnation of the Earth. Helios's sister was the goddess Selene, the Moon. Each day, Helios drove the chariot of the Sun across the sky and each night, he returned to the East through Oceanus, the sea covering the other side of the world. According to Homer, the chariot was driven by four fiery steeds named Pyrois, Aeos, Aethon, and Phlegon.

In the *Odyssey*, Homer relates the story of how Odysseus and his crew arrive at Thrinacia, the island where Hyperion kept the sacred red cattle of the Sun. Odysseus warns his men not to touch the animals, but when supplies run short, they slaughter and eat a few of the cattle. Upon hearing the news, Hyperion warns Zeus (father of the gods) that if the men are not destroyed, he will take the Sun to the underworld. In response, Zeus destroys Odysseus's ship with a lightning bolt, killing all but Odysseus.

As Roman influences filtered into Greek culture, Helios was increasingly identified with Apollo, the god of light. Some literature even conflated the two gods. Though the Romans had their own Sun god, Sol, Apollo slowly became the god most closely associated with the cult of the Sun in both cultures. However, during the same period, Sun worship was on the decline in Greece. In fact, by the late Hellenistic Period, the island of Rhodes was one of the few remaining places where Helios was worshiped fervently. In one religious spectacle, the worshipers in Rhodes would drive a flaming chariot drawn by four horses over the edge of a cliff into the sea.

A fragment of a vase depicting Alexander the Great as the Sun god Helios

JAPANESE: AMATERASU

Amaterasu is the Sun goddess of Shinto, the oldest (and still most practiced) Japanese religion. Some consider Shinto less a religion than a set of rituals to connect the Japanese people to their ancestors. However, the practice is replete with a mythology as complex as any religion.

According to Shinto mythology, Amaterasu fled to a cave when her brother, Susanowo, treated her badly. Closing the entrance with an enormous stone, she caused the world to be plunged into darkness. The Shinto gods arranged to have a party just outside the entrance to the cave, where they placed beautiful jewels and a huge mirror. The celestial goddess Amenouzume performed a bawdy dance, making all the gods laugh. The music, laughter, and dancing drew Amaterasu to open the cave. When she saw her brilliance reflected in the mirror, she was so fascinated that she left the cave and brought light back into the world.

An 1918 illustration shows Amaterasu emerging from her cave after hearing the music and laughter of the other gods.

IRANIAN: MITHRA

Though the Sun god Mithra (also called Mithras) appeared in many civilizations (including the Indian, Babylonian, and Greco-Roman), most scholars agree that Mithra's origins can be traced to Iran, or ancient Persia. The god is known to have been worshiped in Iran as far back as 558 BCE.

The similarities between the story of Mithra and the Christian story of Jesus have led many to refer to Mithra as the "Pagan Christ." Indeed, worshipers of both deities have referred to them variously as the Way, the Truth, and the Light. Mithra has also been described as the Good Shepherd. According to Iranian mythology, Mithra was born in a cave on December 25 and acted as the mediator between god and humanity. Mithra was seen as a benevolent deity, the bringer of health and wellness. Early depictions of Mithra show him riding a chariot of fire drawn by white horses.

This Roman relief carving shows the birth of Mithra (Mitrhas) from the cosmic egg. He holds a dagger and a torch and is surrounded by the signs of the zodiac.

NAVAJO: TSOHANOAI

The Navajo worshiped the god Tsohanoai, who was thought to carry the Sun across the sky on his back every day and hang it on a peg on the west wall of his house at night. According to the Navajo, Tsohanoai had two children, Nayenezgani and Tobadzistsini. They lived with his estranged wife in the Far West. Once the children grew older, they sought out their father, hoping he would join them in their quest to fight off the evil spirits tormenting humanity. On their journey, they met Spider Woman, who gave them two magic feathers to keep them safe. Eventually, the children found Tsohanoai and the god was so happy, he gave them magic arrows to help fight off the evil spirits.

The traditional Navajo home, a hogan, is constructed with the door facing east toward the rising Sun.

NORSE: FREYR

The Norse god Freyr was associated with the Sun, fair weather, and fertility, and is often pictured with a large phallus. He was thought to bring humanity pleasure and peace. In Norse mythology, Freyr was presented as a son of the sea god Njörôr. Freyr rode a boar called Gullinbursti and was said to possess a magic ship that could always catch the wind. When not in use, the ship could be folded together and kept in a pouch.

According to the most extensive Norse myths, Freyr fell in love with the mortal Gerôr. To take her as his wife, Freyr was forced to give up his magic sword, which was capable of fighting on its own. However, according to Norse prophesy, without the sword, Freyr will be killed during Ragnarök, an apocalyptic battle that will end with the world submerged in water and humanity rebirthed by the only two human survivors.

An eleventh-century Swedish statuette of the Norse Sun god Freyr

POLYNESIAN: RA OR LA

In Polynesia, the Sun god was called Ra or La, but he was not revered as much as in other mythologies. In fact, he was forced to slow down in his trip across the sky by the demigod hero, Maui. According to Polynesian mythology, there was never enough time during the day for Maui's mother to make bark cloth, so Maui fashioned a rope from the sacred tresses of his wife, Hina. Then one morning, as the Sun was rising, Maui caught it and beat it with the magic jawbone of his deceased grandmother. After that, the battered Sun could only limp along the sky, creating a longer day for Maui's mother to work.

Besides using his dead grandmother's jawbone to slow down the Sun, Maui used it to fish islands, including New Zealand, out of the sea.

MESOPOTAMIAN: SHAMASH

The ancient Mesopotamians worshiped a Sun god named Shamash. Because Shamash could view everything on Earth, he was known as the god of justice and was depicted as a king seated on a throne. He had two children, Misharu, who represented the law, and Kittu, who represented justice. At night, Shamash was thought to travel through the underworld and exit every morning in the East to begin his journey across the sky to the West. Shamash was often depicted symbolically as a winged solar disk.

Shown seated on a throne, Shamash hands the code of laws to King Hammurabi of Babylon. The laws are carved into the bottom of this roughly 4,000-year-old stela, which is now in the Louvre in Paris.

"THE PURPOSE OF LIFE IS THE INVESTIGATION
OF THE SUN, THE MOON, AND THE HEAVENS."

—ANAXAGORAS, 459 BCE

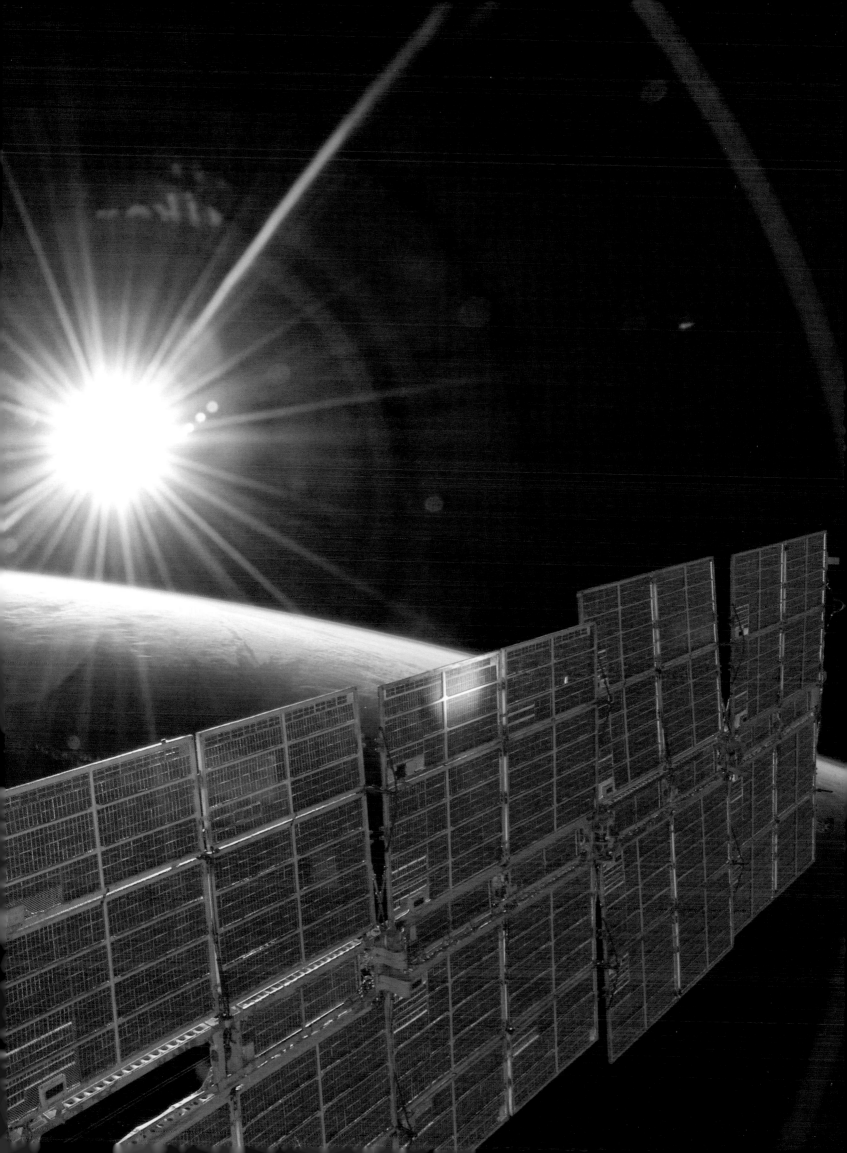

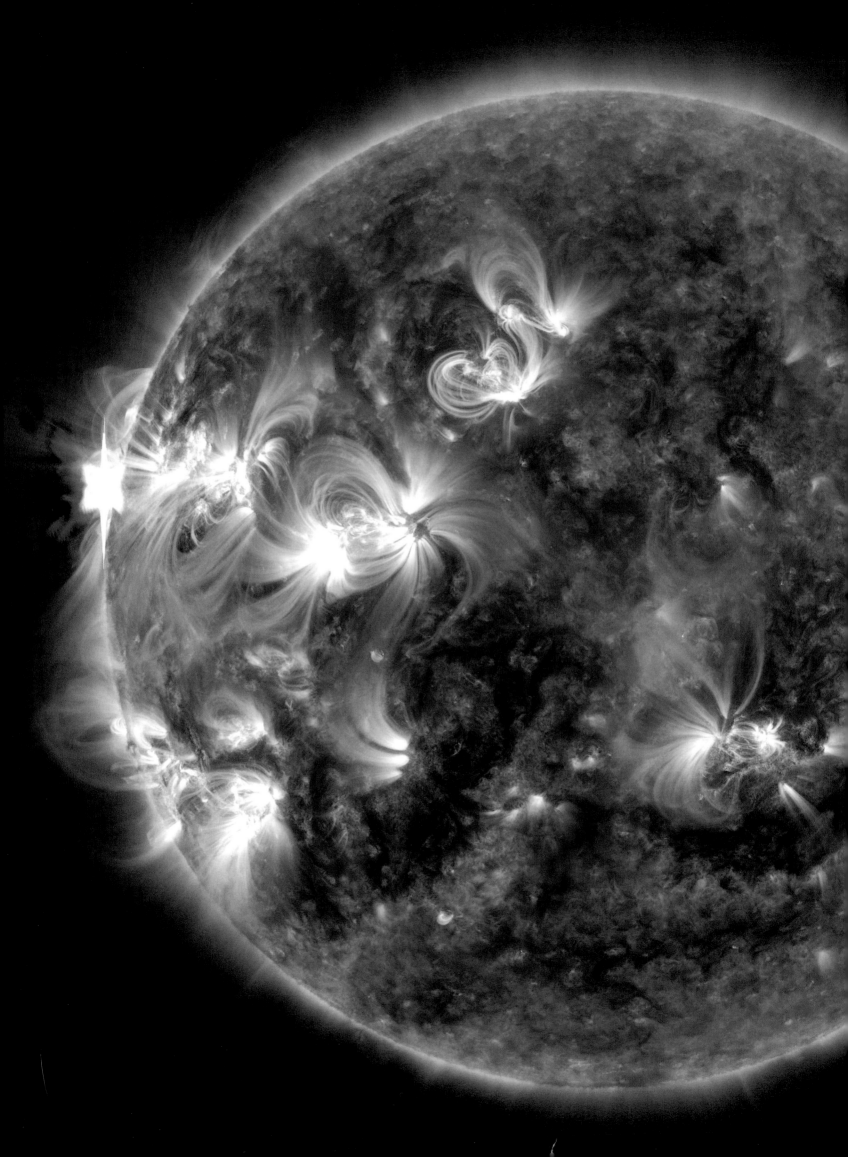

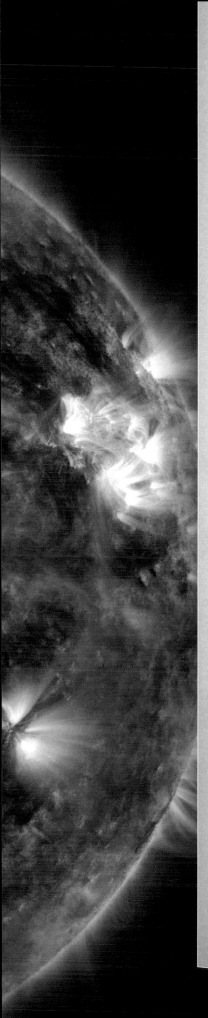

THE HISTORY OF OUR SUN

When early humans observed the sky, they saw that the Sun, stars, and planets appeared to travel around the Earth. From their perspective, moreover, the Earth seemed to be stationary. The logical conclusion was that the Earth must be the immobile center of the universe, around which everything else traveled in perfect circles. Though this geocentric (literally, "Earth-centered") model predated them, Plato and his student Aristotle, Greek philosophers of the fourth century BCE, provided the best logical explanation for this view.

In his famous book *Republic*, Plato describes the cosmos as a series of crystalline spheres surrounding the Earth, one nested inside the other like a set of Russian matryoshka dolls. The spheres were arranged in order outward from the Earth at the center to the Moon, the Sun, Venus, Mercury, Mars, Jupiter, Saturn, and, finally, the fixed stars, which were located on a giant outer shell Plato called the celestial sphere. He imagined that the spheres turned along a Spindle of Necessity, spun by the three Fates.

THE ARISTOTELIAN MODEL

Aristotle had a more mathematical (but no less erroneous) explanation for the movement of the cosmos. The spherical Earth was still at the center of the universe, but all the other celestial bodies were attached to 47 to 56 completely transparent, concentric spheres made of an incorruptible substance he called aether. But according to Aristotle, the natural tendencies of the elements—earth, water, air, fire—explained the placement of substances in the cosmos. Since earth was the heaviest element, it had the strongest movement toward the center. Water, which was less heavy, tended to form a layer surrounding the Earth. Air and fire, being lighter still, had a tendency to move upward, away from the center. Aether was even lighter and explained why the planets, embedded in solid spheres of the substance, were the farthest out.

In addition, Aristotle held that there was one outermost sphere where the "Prime Mover" lived. Aristotle imagined that the Prime Mover caused this outermost sphere to rotate at a constant angular velocity, and that this turning motion was imparted to each inner sphere, causing the whole universe to spin. Aristotle's model allowed for each of the concentric spheres to have a different but constant velocity and, therefore, could explain many features of planetary motion.

Despite the apparent logic of Aristotle's model, it could not account for all of the movement of celestial bodies the ancient Greeks observed. The most problematic of these observations was the apparent retrograde motion of some of the planets. Every so often, it appeared that a few of the planets would slow, start traveling in the opposite direction, slow, and then travel back along the original direction of their orbits. Aristotle's aether spheres, always the same distance from the Earth and moving at constant velocities, could not explain these observations.

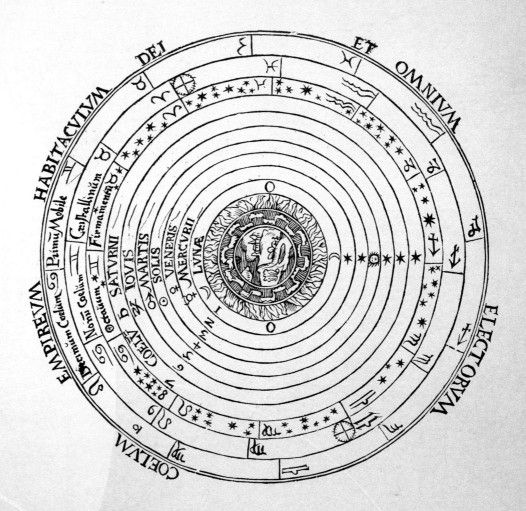

A woodcut showing Aristotle's geocentric system of the universe from the 1539 edition of Cosmographia, *by Peter Apian.*

THE PTOLEMAIC MODEL

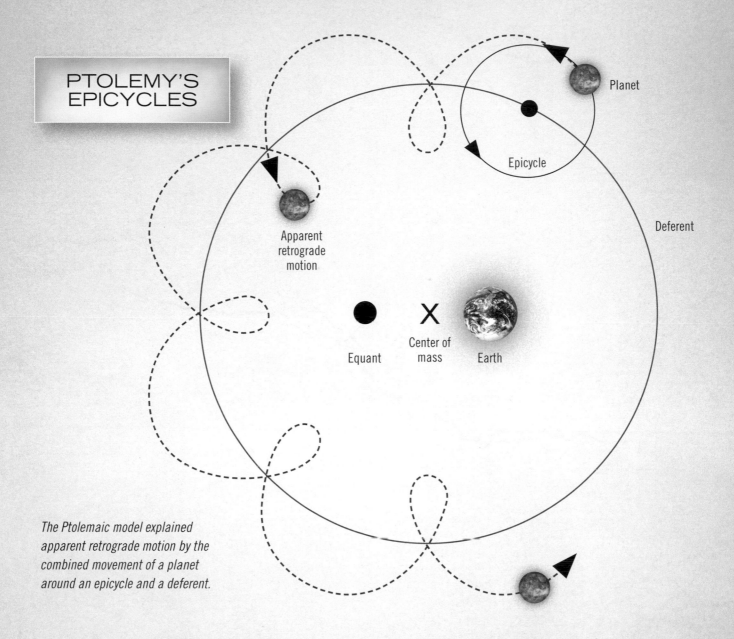

PTOLEMY'S
EPICYCLES

Planet

Epicycle

Deferent

Apparent
retrograde
motion

Equant

X
Center of
mass

Earth

*The Ptolemaic model explained
apparent retrograde motion by the
combined movement of a planet
around an epicycle and a deferent.*

Over 500 years later, in his book *Almagest*, the Hellenistic astronomer Claudius Ptolemaeus devised a model of
the cosmos that kept the Earth at the center of the universe but accounted for the apparent retrograde motion
of the planets. According to Ptolemy (as he was known to his friends), each planet moved along two or more
different spheres. One was called a deferent and the other, an epicycle.

The deferent was a ring centered on a point (the center of mass) halfway between the Earth and a hypothetical
point in space that Ptolemy named the equant. The center point of a planet's epicycle was on the deferent. As
the planet spun around on its epicycle, it simultaneously traveled around the deferent. (Some planets had two
epicycles. The first one was centered on the deferent and the second one was centered on the first epicycle.)
These combined movements caused the planets to sometimes move closer to the Earth and accounted for why
terrestrial observers were tricked into thinking that the planets sometimes slowed, stopped, or even moved in the
opposite direction. This elegant—if complicated—explanation for apparent retrograde motion was so persuasive
that European and Islamic astronomers accepted Ptolemy's cosmological model for over a millennium.

HELIOCENTRISM

Despite the dominance of the Aristotelian and Ptolemaic models, there were several non-geocentric models of the universe proposed by various philosophers, mathematicians, and astronomers as far back as the fifth century BCE. The Greek philosopher Philolaus, for example, speculated that the center of the universe was a central fire around which the Earth, Sun, Moon, and planets all revolved in circular motions.

The first person to have proposed a heliocentric, or Sun-centered, model, however, was the mathematician Aristarchus of Samos in the third century BCE. Although Aristarchus's writings describing a heliocentric model have been lost to time, we have descriptions of them by some of his contemporaries. The mathematician Archimedes' book *The Sand Reckoner*, for example, describes a work in which Aristarchus held that the universe was many times greater than previously thought and that the Sun and stars remained stationary while the Earth revolved around the Sun on a circular orbit.

Aristarchus of Samos, an astronomer who posited a heliocentric universe, is pictured in this woodcut from Liber Chronicarum, *also known as the* Nuremberg Chronicle, *published in 1493.*

Later, in Roman-occupied Europe, there was occasional speculation about a heliocentric model. In the fifth century CE, the North African scholar Martianus Capella opined that Venus and Mercury revolved around the Sun. During the late Middle Ages, Nicole d'Oresme, a French bishop, speculated that the Earth rotated on its axis, while Nicholas of Cusa, a German Cardinal, questioned whether there was any reason to hold that either the Sun or the Earth was the center of the universe.

THE MARAGHA REVOLUTION

In 1259, an Iranian scientist and astronomer, Nasir al-Din al-Tusi, built an observatory in the hills west of Maragha (also known as Maragheh), a city in northwest Iran. The best Islamic astronomers studied at the observatory and began to produce alternative configurations of Ptolemy's model that were more accurate at predicting planetary motions.

In the fourteenth century, one of these astronomers, Ibn al-Shatir, used trigonometry to demonstrate that the Earth was not, in fact, the exact center of the universe. In his book, *A Final Inquiry Concerning the Rectification of Planetary Theory*, al-Shatir introduced a cosmological model that eliminated the need for Ptolemy's equants, in part by moving the Earth slightly from its position in the center of the universe.

Although, strictly speaking, al-Shatir's model did not place the Sun in the center of universe, many historians speculate that it was the inspiration for the Polish mathematician and astronomer Nicolaus Copernicus, who would propose to the Western world an almost identical (but heliocentric) model some 150 years later. However, no one has yet demonstrated definitively that Copernicus knew about al-Shatir's book or the work being accomplished at the Maragha observatory.

This miniature painting from a sixteenth-century manuscript pictures Nasir al-Din al-Tusi at his writing desk in the observatory west of Maragha.

THE COPERNICAN REVOLUTION

In 1543, Nicolaus Copernicus was the first in the Western world to challenge Ptolemy's geocentric model. In *De Revolutionibus Orbium Coelestium* (*On the Revolution of Celestial Spheres*), Copernicus demonstrated that the motion of celestial bodies can be explained without requiring that the Earth be the center of the universe. In the book, Copernicus laid out seven theses that formed the basis of his model:

1. There is no one center of all celestial spheres.

2. The center of the Earth is not the center of the universe.

3. All the spheres revolve around the Sun, which is the center of the universe.

4. The distance between the Earth to the outermost celestial sphere is so large that the distance between the Earth and the Sun is imperceptible in comparison.

5. The motion that appears in the firmament is not the result of the movement of the heavens, but of Earth's movement.

6. What appears as the motion of the Sun derives from the motion of the Earth around the Sun.

7. The apparent retrograde motion of the planets arises not from their motion, but from the Earth's motion.

Copernicus had more or less completed his manuscript around 1533. But, despite urging from many of his contemporaries, he refused to openly publish the book, fearing the backlash his theses might generate. Copernicus's reluctance to publish his theories was, in part, because heliocentrism conflicted with the prevailing views of the Catholic Church and the common understanding of the universe as explained in the Bible. Chronicles 16:30 states that "the world also shall be stable, that it not be moved," while Ecclesiastes 1:5 notes that "the sun also ariseth, and the sun goeth down, and hasteth to his place where he arose." Copernicus knew that his theories would upend not only modern understandings of astronomy, but also the conventional Christian cosmology derived from scriptures like these.

An illustration of the universe according to the Copernican system from Harmonia Macrocosmica *(1660–61) by Andreas Cellarius*

While Copernicus was discovering truths that contradicted the official Catholic cosmology, the Protestant Reformation, led by German theologian Martin Luther, was sparking outright revolution against the Catholic Church. In 1539, Philipp Melanchthon, a German scholar and avowed ally of Luther, arranged for Copernicus to take on a pupil, Georg Rheticus, a young (and likely Protestant) German mathematician. Rheticus studied with Copernicus for two years, after which he wrote his own book, *Narratio Prima* (*First Account*), which outlined Copernicus's model. When Copernicus saw the favorable reception Rheticus's book enjoyed, he finally agreed to publish *De Revolutionibus*. Its printing was overseen by a Lutheran theologian, Andreas Osiander, who completed the first copy on May 24, 1542, the very day Copernicus died. Legend holds that when Osiander placed the book in his hands, Copernicus awoke from a stroke-induced coma, looked upon his life's work . . . and died.

With so many Protestants involved in the publication of Copernicus's work, one would think it would have received a warm welcome from Martin Luther. But, initially, he was one of the most vocal critics of heliocentrism. Luther is said to have denounced Copernicus's theory during a dinner conversation, evoking the biblical story of the Battle of Gibeon in the Book of Joshua. In the story, the Hebrews, with Joshua leading them, were winning a battle against the rebellious Gibeonites at a place known as Beth-Horon. The Hebrews pursued the Gibeonites down the hillsides but feared that their enemies would escape once night fell. So Joshua prayed and commanded the Sun to stand still. Luther (and others) took this story as evidence that the Sun did, in fact, move and could not, therefore, be the center of the universe.

Legend has it that Copernicus received the first copy of his book De Revolutionibus *on his deathbed, as pictured in this nineteenth-century illustration.*

Copernicus's theories were widely opposed in Catholic circles as well. The German Jesuit Nicolaus Serarius was one of the first to condemn heliocentrism as heretical in a manuscript published in 1609. Like Luther, Serarius cited the passage in Joshua describing the Battle of Gibeon (marking, perhaps, the last time the Lutherans and Jesuits would agree on anything for the next 500 years). One of the strongest denunciations came from an Italian Catholic priest, Francesco Ingoli, who wrote an essay in 1616 condemning heliocentrism as "philosophically untenable and theologically heretical."

"AND YET IT MOVES"

The Catholic Church's growing opposition to heliocentrism would play itself out in a series of events culminating in the Italian mathematician and astronomer Galileo Galilei being found guilty of heresy and sentenced to life imprisonment by the Roman Inquisition. In 1609, after hearing about the invention of Dutch perspective glasses, Galileo constructed his own more powerful telescopes. He quickly began astronomical observations and by early 1610, published a small book, *Sidereus Nuncius* (*The Starry Messenger*), about his discovery of mountains on the Moon, moons in orbit around Jupiter, and the star clouds we now know as stellar nebulae. Though Galileo's claims seemed fantastical, they were soon verified by Jesuit astronomers who had obtained telescopes of their own. Nevertheless, some Catholics defended geocentrism on a biblical basis and refused to even look through a telescope.

A nineteenth-century illustration shows Galileo (to the right of the telescope) demonstrating his telescope to the Doge in Venice.

DIALOGO
DI
GALILEO GALILEI LINCEO
MATEMATICO SOPRAORDINARIO
DELLO STVDIO DI PISA.
E Filofofo, e Matematico primario del
SERENISSIMO
GR.DVCA DI TOSCANA.

Doue ne i congreffi di quattro giornate fi difcorre
fopra i due

MASSIMI SISTEMI DEL MONDO
TOLEMAICO, E COPERNICANO;

Proponendo indeterminatamente le ragioni Filofofiche, e Naturali
tanto per l'vna, quanto per l'altra parte.

CON PRI   VILEGI.

IN FIORENZA, Per Gio:Batifta Landini MDCXXXII.

CON LICENZA DE' SVPERIORI.

R. ASTRON. SOC.   19 MAY 39

The title page from the 1632 edition of Galileo's Dialogue Concerning the Two Chief World Systems

In 1613, Cosimo Boscaglia, a professor of philosophy, was conversing with Galileo's patron, Cosimo II de'Medici, Grand Duke of Tuscany. When Boscaglia asserted that Galileo's claims were contrary to scripture, Benedetto Castelli, a Benedictine abbot and former student of Galileo's, immediately jumped to his teacher's defense. When Galileo heard of the heated exchange, he felt compelled to write to Castelli and explain his view of the appropriate way to interpret scriptural passages concerning the motion of celestial bodies, thus unwittingly setting in motion the events that would lead to his condemnation.

In a fiery sermon delivered in late 1614, a Dominican friar named Tommaso Caccini made one of the first public attacks on Galileo. Preaching on Joshua's Battle at Gibeon, Caccini denounced Galileo and all those who would espouse his beliefs. Sometime in early 1615, Niccolò Lorini, Cassini's friend and fellow Dominican, obtained a copy of Galileo's letter to Castelli and sent it to Cardinal Paolo Emilio Sfondrati, secretary of the Roman Inquisition. Word about what Lorini had done soon reached Galileo, along with the news that the Dominican had gone to Rome to denounce him as a heretic. It was late in the year before Galileo's health would allow him to make the journey to Rome (against the advice of his friends), where he hoped to defend himself from any accusations of heresy.

In February 1616, the Qualifiers, a commission of theologians, reviewed the dispute and delivered a unanimous report finding that heliocentrism was "foolish and absurd in philosophy, and formally heretical since it explicitly contradicts in many places the sense of the Holy Scripture." The next day, Pope Paul V ordered that the report be delivered to Galileo along with the command to abandon his heliocentric beliefs or the church would take stronger actions against him. Despite his objections, Galileo accepted the result, but his reputation had suffered a major blow and heleocentrism was formally condemned as heresy.

In 1623, Urban VIII, an admirer of Galileo, was appointed Pope. After his succession to the Papacy, however, Urban was accused by the Spanish Inquisition of being too close to heretical figures like Galileo and too soft on defending the church. Partially to defend himself against this charge, Urban asked Galileo to present arguments for and against heliocentrism—including Urban's own views—in a book Galileo was planning.

GALILEE ABJURE DEVANT LES JUGES DU SAINT-OFFICE LA DOCTRINE DU MOUVEMENT DE LA TERRE

In this nineteenth-century illustration, Galileo faces the Inquisition in Rome. He was convicted of heresy in 1633—a conviction that was not annulled until 1992.

In 1632, *Dialogue Concerning the Two Chief World Systems* was published to immediate public acclaim. Galileo, however, had gone too far. The book was written as a conversation between a scientist and avowed heliocentrist named Salviati and an intellectually inept philosopher named Simplicio (in Italian, "simpleton"). Pope Urban's arguments against heliocentrism were presented in the form of Simplicio's dialogue and were ridiculed and systematically refuted by Salviati.

As you can imagine, the Pope was none too pleased. Within months of its publication, Urban banned the sale of the *Dialogue* and sent the text to a special commission for review. Fearing the loss of the Pope's good graces, many of Galileo's defenders in Rome abandoned him and, in 1633, the astronomer was ordered to the city to stand trial on suspicion of heresy.

On June 22, after a cursory trial, Galileo was found "vehemently suspect of heresy" and required to "abjure, curse, and detest" his heliocentric opinions. He was sentenced to life imprisonment, but the next day the sentence was commuted to house arrest, which remained in effect until Galileo's death. *Dialogue* was formally banned, and publication of Galileo's works forbidden for all time. Legend has it that upon hearing the sentence, Galileo muttered the words "eppur si muove"—and yet it moves.

WHO INVENTED THE TELESCOPE?

An 1863 wood engraving depicts the moment Hans Lippershey gets his idea for a telescope while observing two children playing with lenses in his shop.

Who invented the telescope has been a controversial subject since the seventeenth century. The Dutch lensmaker Hans Lippershey is credited with designing the first working optical telescope. Lippershey was in the lens-making business, an industry that had its genesis with the invention of spectacles in thirteenth-century Italy.

According to one story, Lippershey was watching two children playing with lenses in his shop and observed that they could make a faraway weather vane appear closer by looking through two lenses. Whether inspired by children or something else, Lippershey's first telescope, which was called a "Dutch perspective glass," used two convex lenses to magnify images to three times their size. Lippershey applied to the States-General of the Netherlands for a patent on his instrument in 1608, beating another Dutchman, Jacob Metius, by only a few weeks.

Because there were competing patent claims, Lippershey's application was denied. However, at the time, the Dutch Protestants were at war with the Catholic Spaniards. Believing Lippershey's invention might help in the struggle, the States-General paid him handsomely to construct several of the devices. Afterward, the secret of Lippershey's invention was leaked to the world, leaving no practical way for the lens-maker to claim exclusive rights to his design.

Although Lippershey is most often credited with inventing the telescope, there is some documentary evidence that an Englishman was using a working telescope over 40 years earlier. In the preface to his 1571 book *Pantometria*, Thomas Digges, son of the English mathematician and surveyor Leonard Digges, described a device his father had used sometime between 1540 and 1554:

> My father, by his continual, painstaking practices, assisted with demonstrations Mathematical, was able, and sundry times hath by proportional Glasses duly situated in convenient angles, not only discovered

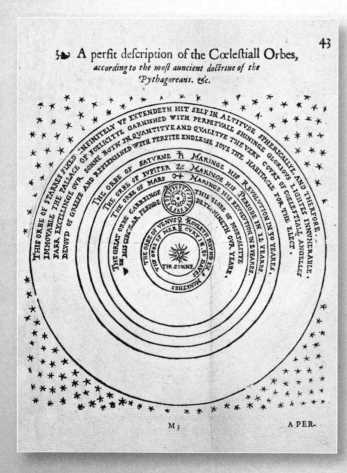

Thomas Digges added this diagram of a heliocentric universe as an appendix to the 1576 edition A Prognostication Everlasting, *by his father, Leonard Digges.*

Galileo presents his telescope to Doge Leonardo Donato in this painting by Henri-Julien Detouche.

things far off, read letters, numbered pieces of money with the very coin and superscription thereof, cast by some of his friends of purpose upon downs in open fields, but also seven miles off declared what hath been done at that instant in private places.

Thomas's writings suggest that Leonard Digges may have created a rudimentary instrument incorporating lenses and a concave mirror. However, if he did, the device was lost after the senior Digges took part in an unsuccessful rebellion in 1554 against England's Catholic Queen Mary, for which he was condemned to death. He only escaped the gallows by forfeiting all of his estates to the crown.

No matter who invented the telescope, the first commercial versions came to market in the Netherlands. Generally, they were composed of a convex and a concave lens together so that the reflected images would not appear inverted. (Try looking at your reflection in a shiny spoon and you will soon see the problem with using only concave lenses.)

In June 1609, Galileo was in Venice when he heard about the new Dutch perspective glasses and their ability to make distant objects appear larger and closer. According to Galileo, upon returning to his home in Padua, he took less than 24 hours to construct his first telescope by fitting a convex lens at one end of a lead tube and a concave lens at the other. A few days later, he completed construction of an even better design, which he took to Venice and presented to Leonardo Donato, the city's chief magistrate. Donato presented it to the city's senate, which was so impressed it doubled Galileo's salary and gave him a lifetime position at the University of Padua.

Johannes Kepler, a German mathematician, analyzed the optics of telescopes mathematically in his 1611 book *Dioptrice*. Kepler determined that using two convex lenses at precise angles would greatly improve magnification but not solve the inverted image problem. Kepler, however, never actually constructed a telescope using this design. It was not until around 1630, when the German Jesuit priest and astronomer Christoph Scheiner constructed a telescope using Kepler's design, that the mathematician's theory was proven in practice. From then on, Kepler's double-convex design became all the rage.

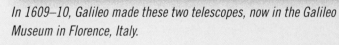

In 1609–10, Galileo made these two telescopes, now in the Galileo Museum in Florence, Italy.

The second model of Isaac Newton's reflecting telescope, built in 1671.

Although the thought of using curved mirrors to create images was known since the fourth century BCE, the earliest known reflecting telescope was not completed until 1668, when Sir Isaac Newton created an instrument using two mirrors made from an alloy of tin and copper. The first mirror reflected an image on a secondary diagonal mirror near the primary mirror's focus point. This secondary mirror then reflected the image at a 90-degree angle to an eyepiece mounted on the side of the telescope. To this day, we call these types of devices Newtonian telescopes.

SPECTROSCOPY: SHEDDING LIGHT ON THE MOVEMENT AND COMPOSITION OF STARS

We have never sent a probe into the Sun. We haven't even come close. If all goes as planned, however, in 2018, NASA will launch Solar Probe Plus, a spacecraft that will come closer to the Sun without vaporizing than any previous human-made object. Even with the most sophisticated heat shielding ever devised, Solar Probe Plus will still need to stay 5.9 million kilometers (3.67 million miles) away from the Sun's chromosphere if it's to be of any use.

Since we have never been to the Sun and can't get close enough to take samples of solar material, you are probably wondering how we can say with any certainty that it is made mostly of hydrogen and helium rather than, say, ice. (If you think that notion is just silly, consider the book that a scientist and theologian named Charles Palmer published in 1798 with the lengthy title *A Treatise on the Sublime Science of Heliography, Satisfactorily Demonstrating that Our Great Orb, the Sun, to be Absolutely No Other than a Body of Ice.* Palmer's theory was that the Sun was a kind of giant comet in the shape of a lens that focused God's energy on the Earth.) The answer is that, through a series of fortunate events, scientists stumbled upon some quirky characteristics of light that tell us an awful lot about where it came from. In fact, in turns out that photons of light are little packets of information, which, if manipulated in just the right way, will reveal the composition and movement of the stars, including our Sun, that created them.

MOVEMENT

Have you ever listened to an ambulance rushing a patient to the hospital? As it gets closer and closer to you, its siren grows higher and higher in pitch until it races past you. Then the siren creeps lower and lower in pitch until the ambulance is out of hearing distance. This phenomenon is known as the Doppler effect, after the Austrian physicist Christian Doppler who first explained it in 1842. The Doppler effect is due to the compression and expansion of sound waves relative to the position of the listener. As the ambulance races toward you, the sound waves from its siren are compressed together. Your ears hear the shorter distance between the waves as a higher pitch. Then, as the ambulance speeds past, the sound waves from the siren are stretched farther and farther apart. As the distance between the waves grows, your ears hear them as a lower pitch.

An 1830 lithograph portrait of Christian Doppler

Doppler correctly predicted that the effect should be produced by all waves, including light from distant stars. If a star were moving closer to the Earth, it should compress the light "waves" seen by someone standing on the planet. But how could such a distortion be detected? The answer lies in spectroscopy, the study of the interaction between matter and the energy it radiates.

SPECTROSCOPY

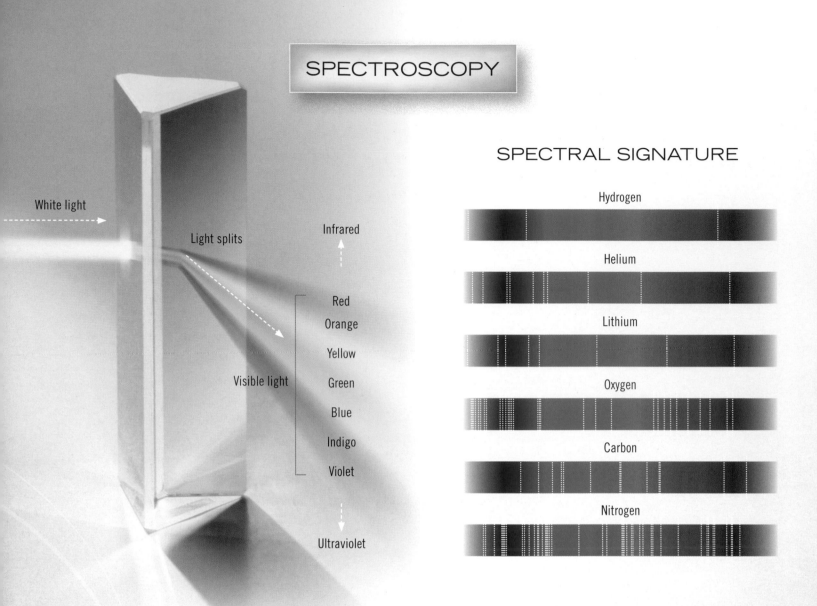

SPECTRAL SIGNATURE

White light

Light splits

Infrared

Red
Orange
Yellow
Green
Blue
Indigo
Violet

Visible light

Ultraviolet

Hydrogen

Helium

Lithium

Oxygen

Carbon

Nitrogen

COMPOSITION

In 1835, the English scientist and inventor of the microphone, Charles Wheatstone, first observed that metals gave off different colors when sparked and could be distinguished by the differences in the spectrum of electromagnetic radiation they emitted. When the electrons in atoms are excited—say, by heating them—the energy they absorb pushes the electrons within each atom into higher energy orbits. When the electrons leave their excited state, they fall back down to a lower energy orbit and release the excess energy in the form of photons. The wavelength (or frequency) of these emitted photons forms the emissions spectrum of the heated element. Every material will emit photons at different frequencies.

In 1854, David Alter, an American physicist, published "On Certain Physical Properties of Light Produced by the Combustion of Different Metals in an Electric Spark Refracted by a Prism," an article that identified the emissions spectra of twelve different metals. Alter speculated that spectrum analysis could be used in astronomy to detect the elements within shooting stars and luminous meteors. A few years after Alter's article was published, the English astronomer Sir William Huggins and his wife, Margaret, used this method to determine that stars were composed of some of the same elements found on Earth.

LINES AND COLORS

Sometime in 1802, an English chemist named William H. Wollaston was observing sunlight through a prism and stumbled upon a curious phenomenon. Attempting to create a clearer image, he first let the sunlight pass through a narrow slit in a piece of metal. What appeared were several thin black vertical lines at various points along the resulting spectrum. Wollaston described them as "divisions between the colors," even though the lines did not fall neatly between the colors at all. The truth was that the solar spectrum had always appeared as a continuum, one color blending into another without clear division.

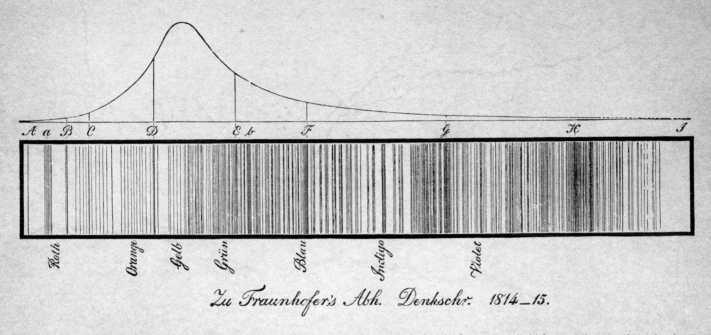

Joseph von Fraunhofer's diagram of the lines of the spectrum, which appeared when he focused sunlight from a telescope through prisms.

About a decade later, a German optician, Joseph von Fraunhofer, was experimenting with various lenses and prisms in an attempt to produce a way of projecting light that was completely smooth. Instead, he found that passing the light through a narrow slit and then through a set of prisms produced a spectrum with dozens of sharp black vertical lines. Through minor tweaks in the configuration of the prisms, he was able to produce a spectrum with 574 of these fixed lines. Cleverly, he named them using letters and numbers (A, B, C1, C2, C3, and so on) starting at the red end of the spectrum. For decades, no one would understand what caused the lines. Yet from that day forward, they would be known as Fraunhofer lines.

In 1859, two professors at the University of Heidelberg, Gustav Kirchoff and Robert Bunsen (who designed the eponymous burner), were—for science or pure entertainment—heating various elements until they glowed and observing the light after it passed through a narrow slit and then a prism. The aspiring pyromaniacs observed that the colors produced by the prism were entirely different from the colors they observed with the naked eye. For example, when the professors burned mercury they saw an eerie blue glow, but what appeared through the prism was a spectrum of violet, green, and yellow. They surmised correctly that the color they observed with their eyes was a mishmash of the three colors produced by the spectrum. The spectroscope, it turns out, allowed them to view the true colors emitted by whatever element they heated.

What's more, Kirchhoff and Bunsen noticed that every element always emitted the same pattern of color, its own unique signature made of light. By burning any substance and viewing the resulting light through the spectroscope, they could determine exactly what elements the substance contained. After experimenting with the process for nearly a year, the two decided to try the spectroscope on the Sun. What they found was a spectrum with the same inexplicable black lines that Fraunhofer had labeled years earlier. Only this time, they figured out what the lines meant. Bunsen excitedly described their discovery in a letter to a friend:

> At present, Kirchhoff and I are engaged in an investigation that doesn't let us sleep. Kirchhoff has made a wonderful, entirely unexpected discovery in finding the cause of the dark lines in the solar spectrum, and he can . . . produce them [in the laboratory] and in the same position as the corresponding Fraunhofer lines. Thus a means has been found to determine the composition of the Sun and the fixed stars.

In the course of burning various substances and viewing them through a crude spectroscope, Bunsen later observed blue spectral lines never seen previously. He deduced that the lines must indicate the existence of a previously unknown element. In the spring of 1860, he was able to isolate the element and named it *caesium*, the Latin word for "deep blue." In 1861, Bunsen used this new process of spectral analysis to discover rubidium, which is derived from the Latin word for "deep red."

A photograph of Gustav Kirchhoff (left) and Robert Bunsen (center) with the English chemist Sir Henry Roscoe

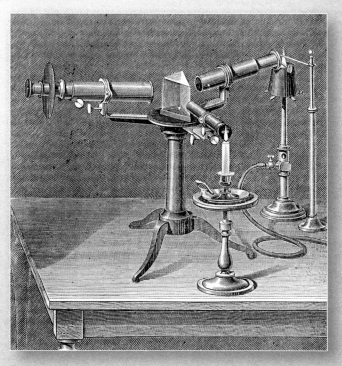

A spectroscope of the type used by Gustav Kirchhoff and Robert Bunsen

RED SHIFT / BLUE SHIFT

Just as Doppler predicted, when we observe light emanating from a star that is moving away from us, its wavelengths will be stretched out. This has the effect of moving the light's radiation signature toward the red end of the electromagnetic spectrum because infrared waves are longer than visible light waves. Consequently, scientists refer to this phenomenon as red shift. Similarly, when we observe light emanating from a star that is moving toward us, its wavelengths will be compressed. This moves its radiation signature toward the blue end of the spectrum because ultraviolet waves are shorter than visible light waves. As you may have guessed, scientists refer to this phenomenon as blue shift.

It turns out that those infuriating Fraunhofer lines have a benefit beyond divulging which elements produced the light. They come in handy when determining red shift or blue shift. All you need to do is compare the spectral signature of an element detected from the light of the star over time. If the lines are shifted to the right (toward the infrared end of the spectrum), the star is moving away from the Earth. If the lines are shifted to the left (toward the ultraviolet end of the spectrum), the star is moving toward the Earth. Even subtle red shifts and blue shifts can tell astronomers about the Sun's movement relative to the Earth—and reveal the startling fact that the Sun does move relative to the Earth all the time, even using a standard astronomical unit.

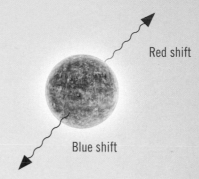

Red shift

Blue shift

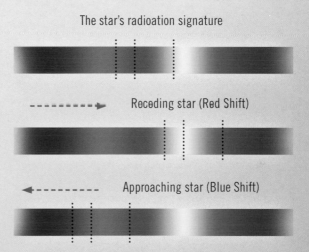

The star's radioation signature

Receding star (Red Shift)

Approaching star (Blue Shift)

Two of the observatories at the summit of Mauna Kea in Hawaii

MODERN TELESCOPES

Not much about the telescope changed for over 220 years after Sir Isaac Newton used a combination of mirrors in his 1668 design. But, as with many things, the dawn of the twentieth century ushered in major changes. In the 1910s, George Willis Ritchey, an American astronomer, teamed with Henri Chrétien, a French astronomer, to invent an entirely new type of telescope. The Ritchey-Chrétien telescope uses two hyperbolic mirrors rather than the combination of convex mirrors and lenses associated with more conventional reflecting telescopes. Today, most professional research telescopes use the Ritchey-Chrétien design.

As telescopes have grown in both size and complexity, scientists have built bigger and more complex observatories to house them. Usually, these observatories are located far from major urban centers to avoid the excess, obtrusive artificial light also known as light pollution. Since the mid-twentieth century, an increasing number of these observatories have been constructed at high altitudes to avoid some of the distortions associated with light filtered through water vapor in the Earth's atmosphere. The largest (and most well known) of these is the Mauna Kea Observatory located near the summit of Mauna Kea volcano on Hawaii's Big Island. Although it is not the highest, Mauna Kea produces the best optical images of any ground-based observatory. The world's highest observatory is the University of Tokyo's Atacama Observatory, located atop a remote mountain in Chile's Atacama Desert at an altitude of 5,640 meters (18,500 feet).

THE FIRST GIANT TELESCOPE

In 1892, François Deloncle, a member of the French Chamber of Deputies, commissioned the construction of a giant telescope as the centerpiece of the Paris Universal Exposition in 1900. It was to be the largest refracting telescope yet constructed, with a lens 1.25 meters (over 4 feet) in diameter and a focal length of 57 meters (over 187 feet), all affixed within a cast-iron tube nearly 60 meters (197 feet) long. Due to its immense size, the telescope had to be mounted in a fixed horizontal position and light from the sky redirected using a movable plane, or siderostat, mirror nearly 2 meters (6.5 feet) in diameter, which would take nine months to grind.

Although the telescope was not intended for scientific use, it could produce images of 500x magnification and more. The French astronomer Charles Le Morvan used it to take several photographs of the surface of the Moon that astonished the readers of Strand Magazine, *which published the photos in the November 1900 issue.*

Unfortunately, its immense size and virtual immobility made the Great Paris Exposition Telescope a hard sell. After the Expo, the company that had built it declared bankruptcy and put the telescope up for auction in 1909. When they could find no buyer, it was broken up for scrap metal. However, the 2-meter (6-foot) siderostat mirror was salvaged and put on display at the Paris Observatory. And in 2007, two of the telescope's lenses were discovered in packing crates in the observatory's basement.

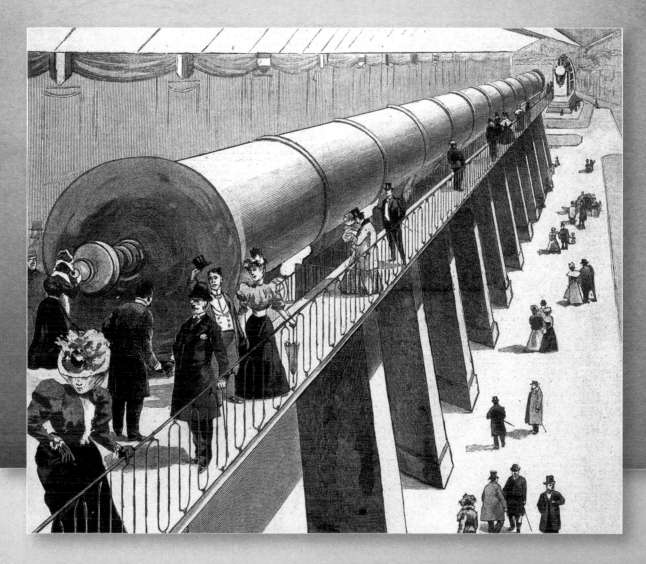

RADIO ASTRONOMY

In the early 1930s, Karl Jansky, an American physicist and engineer working for Bell Telephone, was investigating static interference with transatlantic voice transmissions. Jansky recorded a repeating signal that, based on his calculations, appeared to be coming from the constellation Sagittarius in the densest part of the Milky Way. In 1933, Jansky published his results in an article for *Nature*. Shortly thereafter, an amateur radio operator, Grote Reber, built a parabolic dish in his backyard and conducted the first survey of the sky looking for high-frequency radio waves. Radio astronomy, the detection of radio waves from astronomical bodies, was born.

Karl Jansky and his radio antenna system in Holmdel, New Jersey, in the early 1930s

Radio telescopes in the Karl G. Jansky Very Large Array in New Mexico

Radio telescopes differ from optical telescopes in that they look for radiation at wavelengths longer than the infrared frequency of the electromagnetic spectrum. Many astronomical objects—including the Sun—are observable not only in visible light, but also by the radio waves they emit.

In 1972, the U.S. Congress approved funds to build an observatory of 27 independent radio antennas, each measuring 25 meters (82 feet) in diameter, in the desert just outside of Socorro, New Mexico. Six months later, construction began on the Very Large Array (VLA). Over the next 20 years, the VLA probed the universe and helped discover cosmological phenomena—like black holes, quasars, and pulsars—that once were merely theoretical. In 2012, after more than a decade of upgrades, the VLA was renamed the Karl G. Jansky Very Large Array in honor of the father of radio astronomy. (Despite its depiction in movies like Stanley Kubrick's *2010* and Carl Sagan's *Contact*, the VLA is not used in the search for extraterrestrial intelligence.)

A combined image from the Karl G. Jansky Very Large Array and the Hubble Space Telescope captures spectacular jets of energetic particles shooting out at nearly the speed of light from a supermassive black hole in the core of the Hercules A galaxy.

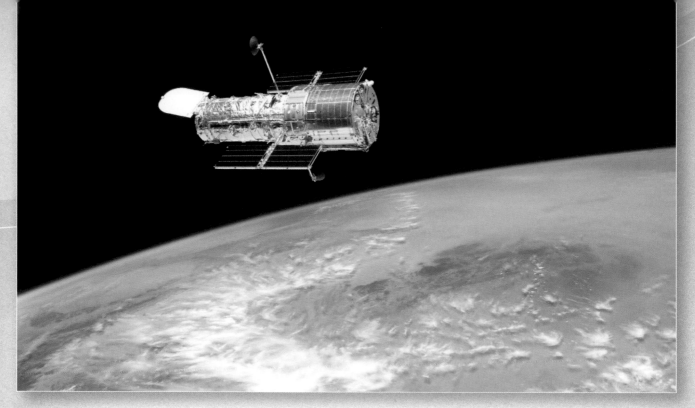

After its fourth servicing mission, the Hubble Space Telescope drifts away from the space shuttle Atlantis *in May 2009.*

SPACE-BASED TELESCOPES

Electromagnetic radiation in the Earth's atmosphere creates distortions (twinkling) for scientists trying to observe celestial objects from the surface of the Earth. In a 1946 paper entitled "Astronomical Advantages of an Extra-terrestrial Observatory," Lyman Spitzer, an American astrophysicist, first conceived of placing an optical telescope in outer space to avoid these distortions. He spent much of the rest of his career trying to bring his idea to fruition. Finally, in 1962, the National Academy of Sciences released a report recommending that the United States develop a space-based telescope as part of its burgeoning space program. Soon thereafter, Spitzer was appointed head of a committee to define the scientific objectives of such a project. Despite his efforts, the European Space Agency would beat the U.S. by launching the Hipparcos space-based optical telescope in 1989.

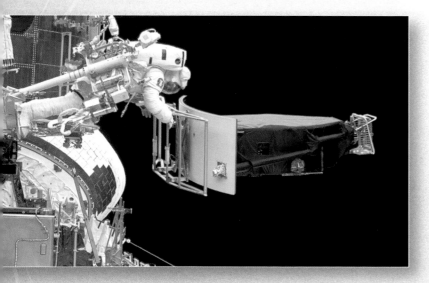

An astronaut removing the Wild Field and Planetary Camera from the Hubble Space Telescope before replacing it with a more powerful instrument during the first servicing mission in 1993.

Though Hipparcos was the first space-based optical telescope, the largest and most versatile is by far NASA's Hubble Space Telescope. Had Hubble been launched in 1983, as originally intended, it would have beaten Hipparcos by almost six years. But technical problems, budget overruns, and NASA's loss of the *Challenger* space shuttle in 1986 caused Hubble to be significantly delayed. When it finally was launched in 1990, scientists discovered that its main mirror contained design flaws that distorted its images. In 1993, NASA shuttle astronauts performed the first service mission of its kind to replace and repair Hubble's components.

Since its repair, Hubble's detailed images of distant objects have led to major breakthroughs in astronomy and in our understanding of the universe. Data from Hubble has formed the basis of over 9,000 academic papers published in peer-reviewed journals. The telescope has helped refine estimates of the age of the universe and uncovered the remarkable fact that the universe, despite everything we thought we knew about gravity, appears to be expanding at an accelerated rate.

A three-light-year-tall mountain of gas and dust in the Carina Nebula taken by the Hubble Space Telescope.

Above: *In 2004, the Hubble Space Telescope captured the swirling interstellar dust given off by V838 Monocerotis, the red supergiant star shining out from the center.*

Opposite page: *This 2013 image of the Small Magellanic Cloud combines data from the Hubble Space Telescope, the Chandra X-ray Observatory, and the Spitzer Space Telescope.*

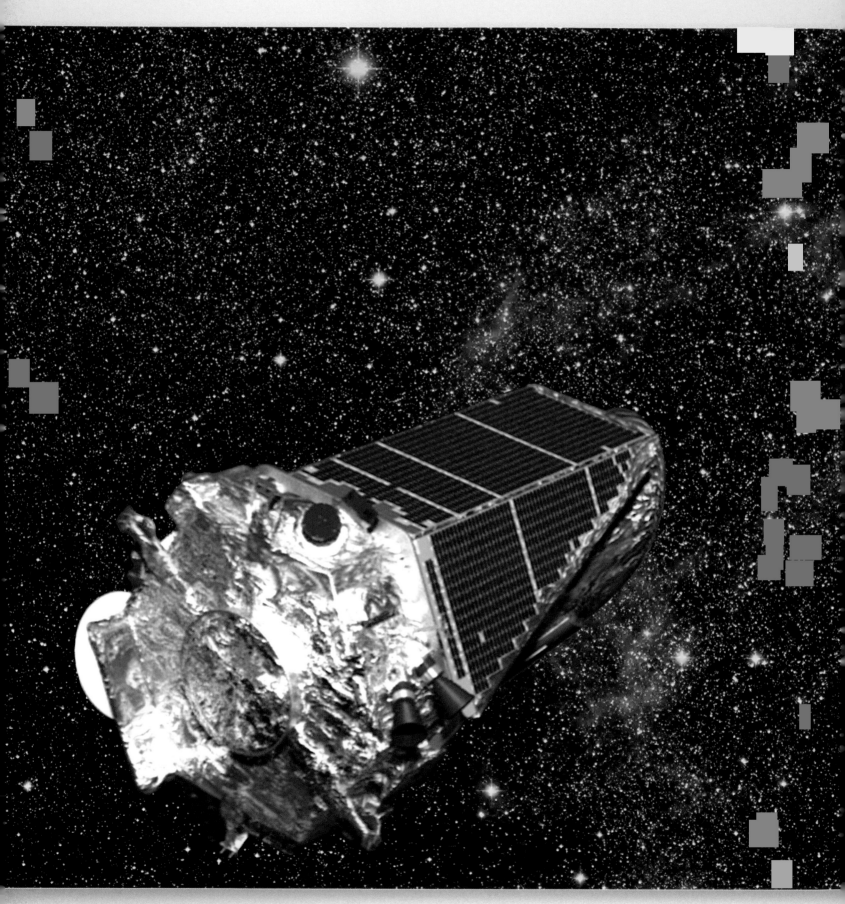

An artist's rendition of the Kepler spacecraft

In March 2009, even while Hubble continued to supply an endless catalogue of breathtaking images of the universe, NASA launched the Kepler Space Observatory, named in honor of Johannes Kepler, the seventeenth-century German mathematician, astronomer, and staunch defender of the Copernican model. While Hubble was designed to look throughout the universe, Kepler's mission is limited to discovering Earth-like planets orbiting other stars. This task is not as easy as it might seem. Just as the brilliance of the Sun obscures objects nearby, it is nearly impossible (even with Kepler's advanced optics) to detect exoplanets, planets outside of our solar system, by looking for them directly. Instead, Kepler uses an advanced photometer, a device that measures brightness, to continuously monitor over 145,000 stars in a fixed field of view. Kepler transmits this data to Earth where scientists analyze it looking for any signs of periodic dimming caused by an exoplanet crossing in front of its host star. As of 2013, Kepler had, through these tiny variations in brightness, detected over 2,165 possible exoplanets, 122 of which have been confirmed.

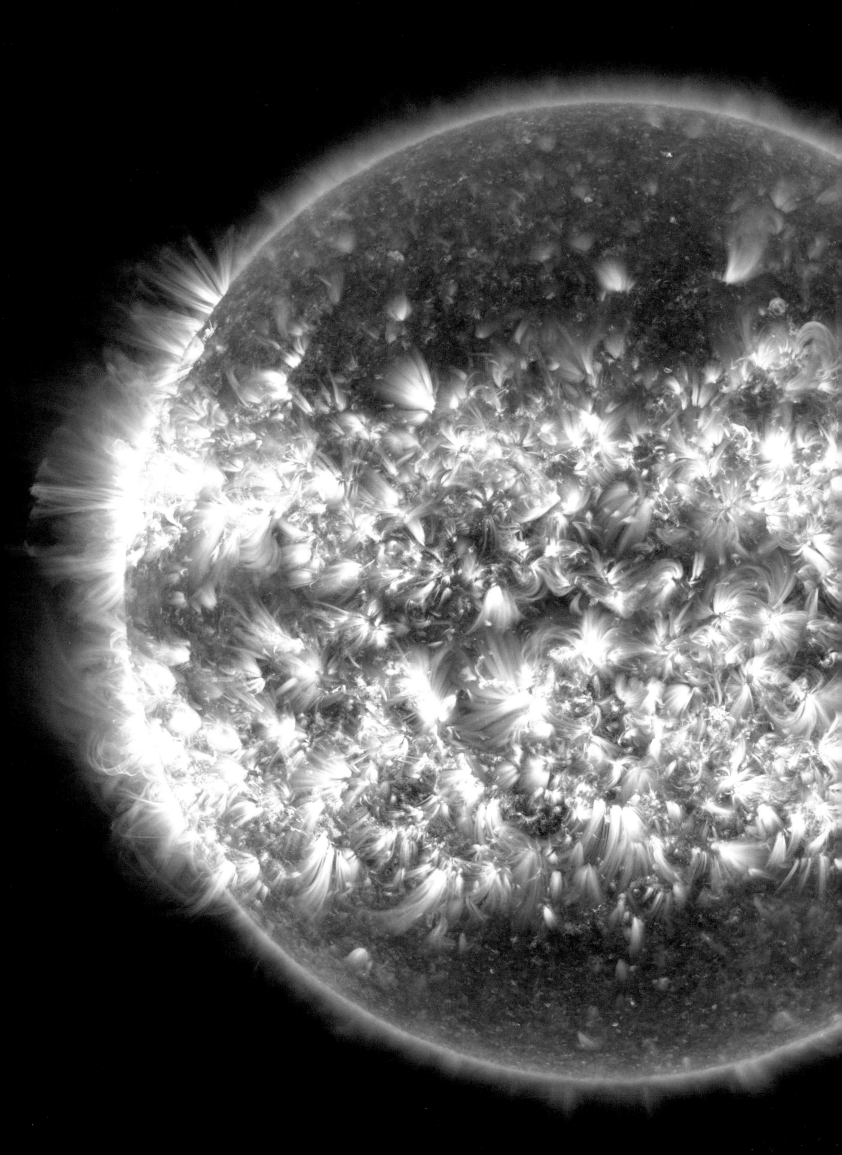

6

THE POWER OF OUR SUN

Scientists know that plasma in the Sun's corona absorbs so much energy that charged particles escape the Sun's gravity and fly out in all directions as a solar wind with speeds that can exceed 1.6 million kilometers (about 1 million miles) an hour. Only recently have they discovered the likely cause. Braids of electromagnetically charged plasma stretching from the chromosphere to the outer reaches of the corona release tremendous amounts of energy as they pull against magnetic field lines, which are constantly trying to straighten the braids on the Sun's surface. The particles emanate in all directions, including straight toward Earth.

THE EARTH'S MAGNETOSPHERE

Fortunately, Earth produces its own electromagnetic field. Earth's inner core is a solid sphere of iron-nickel alloy about 1,220 kilometers (760 miles) in diameter. Its outer core, which is about 2,260 kilometers (1,404 miles) thick, is mostly the same, except that these electrically conducting metals are in a liquid state at temperatures up to 6,100 degrees Celsius (over 11,000 degrees Fahrenheit). Just like the Sun's convection zone, the molten metal in the Earth's outer core creates convection currents as heat is transferred outward toward the mantle. The movement of this liquid metal generates the magnetosphere, an electromagnetic field that extends out from the surface of the Earth.

Scientists began to understand the power of Earth's magnetosphere when the first satellites were launched in the late 1950s. By analyzing data from the satellites, they realized that Earth was surrounded by a belt of highly charged particles stretching from about 50 to nearly 1,000 kilometers (31 to 620 miles) above the surface, in an area known as the ionosphere. Scientists realized that the charged particles were being trapped in an area now called the Van Allen radiation belt by electromagnetic forces emanating from inside the Earth. Essentially, these particles extended Earth's magnetic field much farther into space than anticipated.

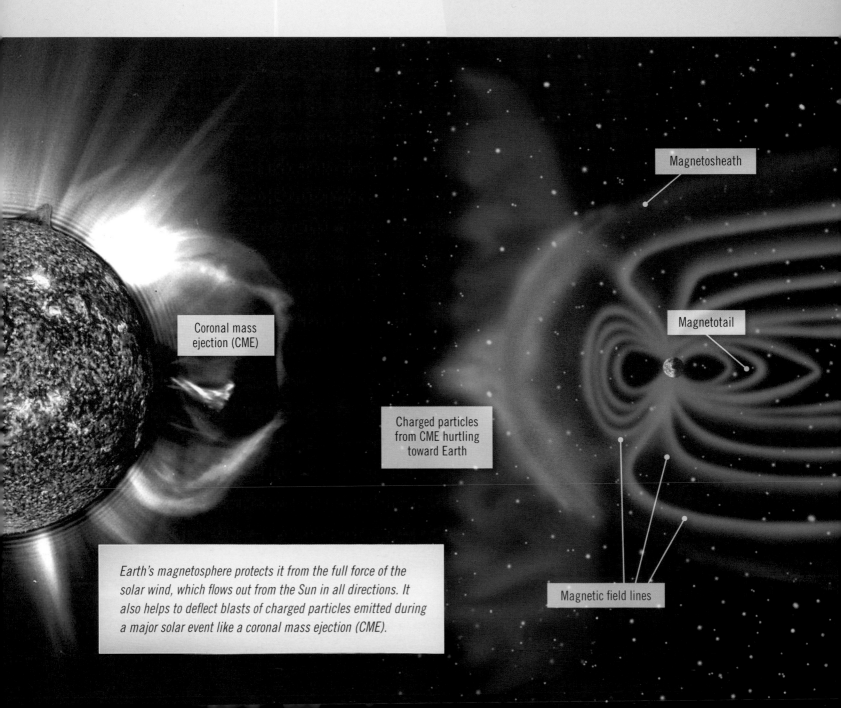

Magnetosheath

Coronal mass ejection (CME)

Magnetotail

Charged particles from CME hurtling toward Earth

Magnetic field lines

Earth's magnetosphere protects it from the full force of the solar wind, which flows out from the Sun in all directions. It also helps to deflect blasts of charged particles emitted during a major solar event like a coronal mass ejection (CME).

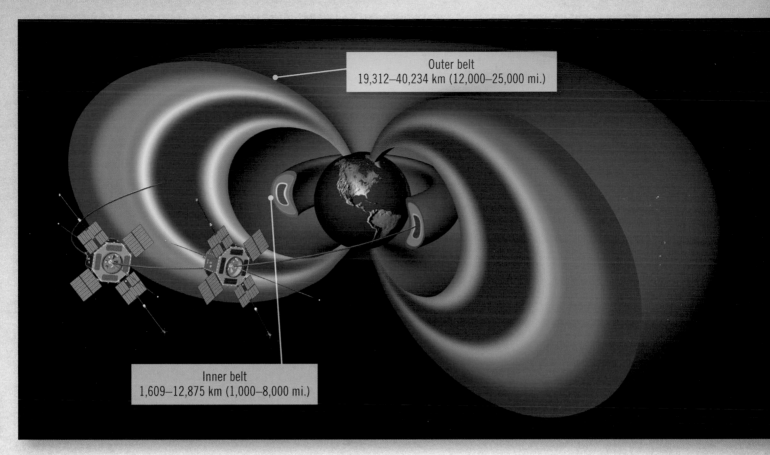

Outer belt
19,312–40,234 km (12,000–25,000 mi.)

Inner belt
1,609–12,875 km (1,000–8,000 mi.)

The Van Allen Probes are two identical spacecraft that were launched in August 2012 to gather detailed data on the Van Allen radiation belts.

The shape of the Earth's magnetosphere depends on its alignment with the solar wind. Pressure from the solar wind compresses the magnetosphere on the side of Earth facing the Sun to a distance of about six times the radius of the planet. Since Earth has a radius of 6,371 kilometers (3,959 miles), this means its magnetosphere stretches more than 38,000 kilometers (almost 24,000 miles) out into space. The magnetosphere on the side facing away from the Sun is dragged away from the Earth as the solar wind wraps around the planet. This effect creates what is known as the magnetotail, which stretches as far as 1,000 times the Earth's radius, or some 6.4 million kilometers (almost 4 million miles).

MAGNETIC POLES

The Earth's magnetic field contains lines of electromagnetic charge that dive into the Earth's surface at the magnetic North and South Poles. As you probably know, the magnetic poles are not the same as the geographic North and South Poles. But perhaps you didn't know that the magnetic poles travel around. Because the molten metal sloshing about the Earth's outer core refuses to stay in one place, the magnetic field lines it produces move along with it. In the last hundred years or so, magnetic north has moved nearly 1,046 kilometers (650 miles).

What is really interesting, however, is that, since 1989, its movement has accelerated from about 8 to over 75 kilometers (5 to 47 miles) a year—or almost 5 meters (about 16 feet) every hour! No one knows for certain what is causing the acceleration or if the shifting magnetic pole intends to slow down. Even odder, the movement of Earth's magnetic South Pole, already some 2,900 kilometers (1,800 miles) from the geographical South Pole, is slowing down. While magnetic north is speeding away from geographic north, the movement of magnetic south has slowed to about 5 kilometers (roughly 3 miles) per year.

While over the eastern North Atlantic in March 2012, a crew member on the International Space Station took this spectacular photo contrasting the northern lights on the left with the bright sunrise on the right. The earthly lights are from cities in Ireland and the United Kingdom.

On January 8, 2008, a chance alignment of Earth and Mars allowed NASA scientists to compare how the solar wind impacts the Earth, which has a protective magnetosphere, and Mars, which does not. The results were sobering. While the solar wind created similar pressures on each planet, Mars's atmosphere lost oxygen at ten times the rate of Earth's. Over millions of years, the scientists concluded, the solar wind had stripped Mars of its atmosphere, but the shielding effect of Earth's magnetosphere protected our atmosphere from a large part of the Sun's electromagnetic current.

In 2008, data from NASA's Interstellar Boundary Explorer (IBEX), a small satellite designed specifically to study the solar wind, uncovered exactly how this protective mechanism works. Every planet is drained of atmosphere when the passing solar wind exchanges electrical charges with charged particles surrounding the planet. Fortunately, in Earth's case, its magnetosphere ensures that the electrical exchange happens in the outer reaches of the atmosphere, where air densities are so low that the interaction of electric currents is relatively weak, at least most of the time.

By-products of the current exchange between the solar wind and the Earth's magnetosphere are the spectacular light displays at each of the Earth's poles: the aurora borealis (northern lights) and the aurora australis (southern lights). The collision of charged particles high above the poles creates stunning colors, which depend on the amount of energy the atmosphere absorbs.

The aurora and underlying atmosphere of Saturn's north polar region, as captured in wavelengths of infrared light by NASA's Cassini spacecraft

When energy from the solar wind collides with ionized atoms in the atmosphere (usually oxygen or nitrogen), electrons in the atoms become excited and momentarily orbit in a higher energy state. When the electrons return to a less excited orbit, the atoms release photons of different wavelengths depending on the nature of the element. Photon emissions from oxygen atoms, for example, appear as green- or rust-colored, while photon emissions from nitrogen atoms look blue or red. The mixture of these colors at different levels of the atmosphere (where elements mix in different proportions) results in a dazzling rainbow display in the night sky.

Usually, these breathtaking light shows are limited to the auroral zones (typically, a ring between 3 and 6 latitudinal degrees thick at about 10 to 20 degrees from the magnetic poles). However, during particularly powerful magnetic storms (for instance, when a highly charged coronal mass ejection smacks directly into Earth's atmosphere), auroras can be seen as far south (or north) as the tropics.

Earth is not the only planet to produce auroras. Jupiter and Saturn both have magnetic fields that are far more powerful than those on Earth. Jupiter's magnetic field, for example, is more than 14 times stronger than Earth's. NASA's Hubble Space Telescope has taken pictures of massive and thrilling auroras around both planets, as well as around Uranus and Neptune.

A HOLE IN THE MAGNETOSPHERE?

On February 17, 2007, NASA launched five identical satellites, each weighing about 129 kilograms (about 285 pounds), into orbit within Earth's magnetosphere. The Time History of Events and Macroscale Interactions During Substorms (THEMIS) mission was designed to analyze how energy from the magnetosphere is released during substorms, sudden disturbances in Earth's magnetic field that send electric charges to the outer edge of the ionosphere. Later that year, THEMIS discovered that Earth's upper atmosphere is connected directly to the Sun's corona via "ropes" of electromagnetic energy known as Birkeland currents, named for Kristian Birkeland, the Norwegian scientist who first elucidated the nature of the northern lights in 1908.

But the Birkeland currents were far from THEMIS's most surprising discovery. On June 3, 2007, sensors on board all five satellites recorded a wave of charged solar particles streaming into the magnetosphere at a rate of 10 octillion (that's 10 followed by 27 zeros) particles per second, far more than normal. Buffeted by this torrent of particles, high above the poles of the Earth, where the Birkeland currents create conduits between the Earth's magnetosphere and the Sun's corona, the currents suddenly expanded, creating a breach in the magnetosphere almost four times the size of the Earth itself!

What was most surprising, however, was that a hole was ripped open in the magnetosphere even when the solar magnetic field and the Earth's magnetic field were aligned. Scientists had always believed, for example, that when a gust of northward-oriented solar magnetism came along and slammed into the Earth's magnetosphere above the equator (where magnetic field lines also point north), the two fields would reinforce one another, strengthening that part of the Earth's magnetic shield against the charged particles of the solar wind. Instead, THEMIS revealed that northern-aligned magnetic currents could tear holes in the magnetosphere at the poles, loading the protective layer of the outer atmosphere with tons of charged plasma and setting up conditions for the perfect solar storm, should a coronal mass ejection or other disturbance hit the Earth at just the right moment.

An artist's concept of the THEMIS spacecraft as they might appear in orbit.

ULTRAVIOLET RADIATION

Just beyond the violet end of the visible light spectrum, the wavelength of photons shortens and the electromagnetic spectrum enters the ultraviolet (UV) spectrum. More commonly known as UV radiation, energy at this end of the spectrum is most closely associated with X-rays and, therefore, raises serious concerns about cancer. The UV end of the electromagnetic spectrum usually is subdivided based on the decreasing size of the wavelengths of energy: near ultraviolet (400 to 300 nanometers, or nm), middle ultraviolet (300 to 200 nm), far ultraviolet (200 to 100 nm), and extreme ultraviolet (below 100 nm). (For comparison, a human hair is about 100,000 nanometers wide.) The wavelengths most likely to affect humans are called ultraviolet A (UVA), with wavelengths of 400 to 315 nanometers; ultraviolet B (UVB), with wavelengths of 315 to 280 nanometers; and ultraviolet C (UVC), with wavelengths of 300 to 100 nanometers. Radiation below 10 nanometers is commonly known as X-rays.

In October 2010, a magnetic filament longer than the distance from the Earth to the Moon cut across the Sun's southern hemisphere. The bright spot just north of the filament's midpoint is UV radiation from the sunspot where the filament appears to be rooted.

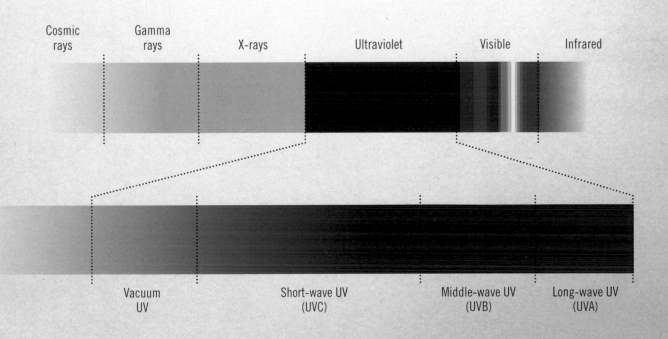

Cosmic rays — Gamma rays — X-rays — Ultraviolet — Visible — Infrared

Vacuum UV — Short-wave UV (UVC) — Middle-wave UV (UVB) — Long-wave UV (UVA)

While the Sun emits radiation at all wavelengths, very little of the UV energy that reaches the Earth's surface has smaller wavelengths than UVC. In fact, even when the Sun is directly overhead, UV radiation makes up less than 3% of the total amount of sunlight that hits the Earth. At the very top of our atmosphere, the total amount of energy absorbed as UV radiation is about 140 watts per square meter (13 watts per square foot). By the time the light has reached the surface, however, the atmosphere has filtered out more than 75% of that energy. Every square meter of the Earth's surface, therefore, is continuously hit with about 34 watts of UV radiation. For comparison, a typical iPhone charger produces about 5 watts of energy.

Though some exposure to UV radiation can be beneficial, overexposure to UVB radiation can cause serious sunburn and has been linked to skin cancer. In fact, the incidence of skin cancer varies by altitude and by latitude. In other words, even on the Earth's surface, exposure to UV radiation will vary wildly depending on how close you are to the Sun. Someone living in Virginia or North Carolina has about twice the risk of developing skin cancer as someone living in Vermont or New Hampshire. The risk doubles again for people living in sunny Florida or Texas.

The incidence of skin cancer in the United States has been rising by about 5% each year. Many doctors attribute this increase to the migration of America's aging baby boomers to warmer climes—like Phoenix, Las Vegas, and Fort Lauderdale—where they are spending more of their retirement years playing golf, tanning on the beach, and generally exposing themselves to more of the Sun's damaging radiation. A 5% annual increase is serious cause for concern. Even at the present rates, nearly 3 men and 1 woman in every 100 will develop melanoma in his or her lifetime.

UVC radiation, an even more potent form of energy, can be quite dangerous, causing the kinds of damage to human DNA that has been linked to up to 92% of all melanomas. In the past, exposure to UVA radiation was considered harmless. In 2011, however, the World Health Organization classified all wavelengths of UV radiation (including UVA) as Group 1 carcinogens, the highest and most dangerous designation for substances known to cause cancer in humans.

The rising rate of skin cancer may be due, at least in part, to the number of retired baby boomers moving to sunnier climates and spending more time outdoors.

Sunscreens help prevent damage to DNA from UVB radiation. In fact, sun protection factor (SPF) levels are also called UVB-PF, for UVB protection factor. SPF levels tell consumers virtually nothing, however, about a sunscreen's ability to protect against UVA exposure (and do not protect against UVC at all). Unless the sunscreen contains compounds like titanium oxide, zinc oxide, or avobenzene, it is doubtful they offer much protection against UVA radiation.

The damaging effects of exposure to UV radiation raises serious health concerns. But if not for exposure to UV radiation, life on Earth may not have evolved into what it is today (or at all). Many evolutionary biologists believe that exposure to the Sun's UV radiation is what allowed early forms of life to evolve the kinds of reproductive proteins and enzymes that were necessary to create life-forms capable of living on the surface of the planet.

The theory is that, as early unicellular life-forms rose to the surface of primordial oceans, they were exposed to lethal amounts of UV radiation. (This was long before the Earth had developed an ozone layer that helps block out much of the radiation.) Most of these organisms died as a result. The few that survived were those with genetic mutations that made them produce special enzymes capable of overcoming the cellular disruption caused by exposure to UV radiation. Survival of the fittest of these organisms eventually led to the evolution of organisms capable of surviving on the surface of the planet, in full exposure to UV.

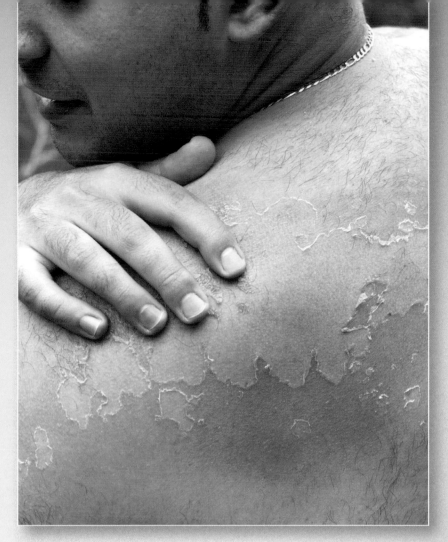

According to the Skin Cancer Foundation, the risk of developing melanoma doubles for a person who has had sunburn five or more times.

Only sunscreens that contain compounds like titanium oxide, zinc oxide, or avobenzine help protect against UVA as well as UVB radiation.

VITAMIN D: THE SUN'S MIRACLE SUPPLEMENT

Strictly speaking, vitamin D is not a vitamin at all. Usually the term *vitamin* is reserved for essential organic compounds that organisms cannot create on their own but must obtain through their diets. Most mammals, however, have evolved to synthesize vitamin D when exposed to sunlight. In fact, scientists believe that land-based animals have been making their own vitamin D since the early Carboniferous period, some 350 million years ago. Though we have evolved into very different creatures since then, the way our bodies produce vitamin D has remained largely the same.

In only 30 minutes of whole-body exposure to sunlight, your body can produce more vitamin D through a photochemical process than most supplements can supply in pill form. When the outermost layer of the skin is exposed to UVB, a ring of 7-dehydrocholesterol's elemental structure breaks open to form previtamin D_3. This is then transformed by the skin's heat into the secosteroid vitamin D_3, a precursor to vitamin D.

Carrier proteins in the bloodstream transport the vitamin D_3 from the skin to the liver, where special enzymes convert it to calcifediol. Calcifediol is a biologically inactive substance, a form of stored vitamin D that has the potential to be converted into active vitamin D. In our bodies, calcifediol is converted to the active form of vitamin D in two different ways.

Most of the calcifediol goes to the kidneys, where more enzymes convert it into active vitamin D. Then the kidneys attach the vitamin D to special carrier proteins that help transport the vitamin to organs throughout the body. There is some evidence that, as it circulates, vitamin D helps regulate the concentration of calcium and phosphate in the blood, which is essential for healthy bone growth.

Calcifediol can also be converted into active vitamin D in macrophages, special immune cells that literally "eat" germs in your body. The microphages use the vitamin D locally, usually to reduce inflammation and help the body defend itself against germs trying to invade a particular area.

A half-hour of sunbathing can produce more vitamin D than most daily supplements.

UVB radiation

Skin

7-dehydrocholesterol

HO

Previtamin D₃

HO

Vitamin D₃

HO

HOW THE BODY MAKES VITAMIN D

Energy in the form of photons breaks open the structure of 7-dehydrocholesterol (a type of cholesterol found in the skin), allowing our bodies to transform it into an essential nutrient.

Vitamin D₃ enters the bloodstream

Liver converts vitamin D₃ to calcifediol

OH

Calcifediol

HO

Microphages convert calcifediol to active vitamin D, which is used locally to reduce inflammation

Kidneys convert calcifediol to active vitamin D, which carrier proteins transport to organs throughout the body

OH

Active vitamin D

HO   OH

OH

Active vitamin D

HO   OH

VITAMIN D AND AUTISM

Autism is the fastest-growing developmental disability on Earth, with the number of cases growing by over 1,100% since 1990, according to one California-based study. Although there is no definitive scientific evidence available as yet, researchers at the Vitamin D Council believe these soaring rates are linked to lack of exposure to sunlight. They point to statistics showing higher rates of autism in areas with more cloud cover as well as the fact that Black people who live in higher latitudes and whose vitamin D levels are about half those of Caucasians living in the same areas have nearly twice the rate of autism. But in lower, sunnier latitudes, where people tend to spend more time outdoors, autism among Blacks is rare.

Doctors have long regarded vitamin D as beneficial to human health. Lack of vitamin D can cause osteomalacia (commonly known as rickets), a disease that can lead to softening of the bones and impede healthy growth, especially in children. Low levels of vitamin D have also been associated with multiple sclerosis and some forms of cancer.

Recently, some doctors have extolled the benefits of vitamin D on the body's ability to fend off infections from influenza and tuberculosis. The Vitamin D Council, a nonprofit educational organization, has suggested that the zeal to prevent UV exposure, coupled with an increasingly sedentary culture that prefers staying indoors in front of a computer screen to playing in the sun, is causing chronic deficiencies in naturally produced vitamin D. One report from the Centers for Disease Control (CDC) estimated that 90% of infants in the United States are not getting the recommended amounts of vitamin D.

Scarier still, some studies suggest a link between vitamin D deficiency and the growing rates of autism. In 2010, the National Academy of Sciences issued a report raising the recommended daily intake (RDI) of vitamin D to 400 IU (international units, a measure of vitamin potency) for infants, 600 IU for adults, and 800 IU for the elderly. This boost, although a significant increase from previous recommendations, was too conservative for four of the Academy's physicians, who resigned in protest. Those physicians felt higher levels (up to 10,000 IU for adults) were called for.

Whatever the dosage, increasing numbers of studies support the benefits of steady, naturally produced, and higher levels of vitamin D. According to Bob Berman, one of the world's foremost experts on the Sun, sufficient exposure to sunlight could prevent as many as 150,000 cancer deaths per year in the United States alone. It may also reduce infection rates and improve the health and well-being of millions of people suffering from bone and cardiovascular diseases.

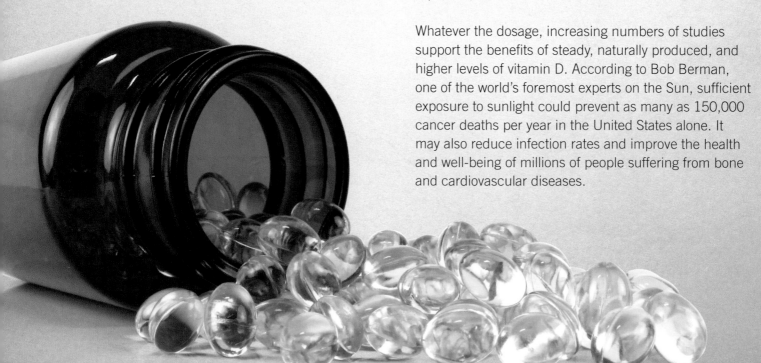

SOLAR POWER: HARNESSING THE SUN'S ENERGY

At any given time, about 174,000 terawatts of energy in the form of solar radiation is blanketing the Earth. Earth's upper atmosphere reflects about 30% of that energy back into space. Some of it is further reflected or absorbed by clouds closer to the Earth's surface. Still, the total amount of solar energy that makes it to the surface is enormous. Each year, more solar energy reaches the surface of our planet than nearly twice the energy we can squeeze out of the planet's entire supply of coal, oil, natural gas, and uranium.

Since prehistoric times, humans have found ways to harness the Sun's energy for light and heat. But only in the last hundred years or so have we developed ways of collecting relatively large amounts of solar energy and converting it to electricity. Several different methods for generating electricity from solar energy have been developed since humans first discovered that sunlight could power an electric current. But, for all practical purposes, there are three basic methods of generating large amounts of solar power: solar photovoltaic (PV) arrays, concentrated solar power (CSP) arrays, and solar power towers.

SOLAR PHOTOVOLTAIC ARRAYS

Solar photovoltaic (PV) arrays generate electricity from sunlight using PV cells made of materials whose electrons become excited in the presence of sunlight and produce an electric current. Today, PV cells are made mostly from different types of silicon. You may be surprised to learn that pure silicon is not particularly useful in making PV cells. Instead, manufacturers use a process called "doping" to add certain impurities that make the silicon more conductive.

Silicon doped with phosphorus or arsenic is called N-type silicon, because those impurities cause the silicon to have an excess of negatively charged electrons. P-type silicon, which is doped with boron or gallium, has electron vacancies that create a latticework of positively charged "holes" waiting to be filled with any electrons that might wander by.

When the two types of silicon are brought into close contact, the free electrons in the N-type silicon rush to fill the holes in the P-type silicon. The buildup of combined electrons and holes around the P-N junction (the boundary between the silicon layers) creates an electric field that overlaps the two layers. This electric field acts as a barrier preventing more free electrons from crossing over from the N side to the P side.

Enter sunlight. When photons of the right wavelength strike the PV cell, their energy is absorbed, splitting up electron-hole pairs. Though the electric field allows freed electrons to cross from the P side to the N side, it stops them from moving in the opposite direction. In an effort to reconnect with a hole on the P side, the electrons will follow an external path. An electrode, or electrical contact, at the front of the N layer, channels the electrons into a circuit that connects to an electrode on the back of the P layer. This creates an electric current that, along the way, can be used to power anything from a digital calculator to an entire town.

Dozens of PV cells can be arranged on solar panels. Thousands of solar panels can be arranged into massive arrays that generate enormous amounts of electricity. In 2012, the U.S. Army installed thousands of solar panels at the White Sands Missile Range in the middle of the New Mexico desert, creating the world's largest low-concentration PV array, generating over four megawatts of electricity. (One megawatt, if it can be sustained, is about enough electricity to power a large housing development.)

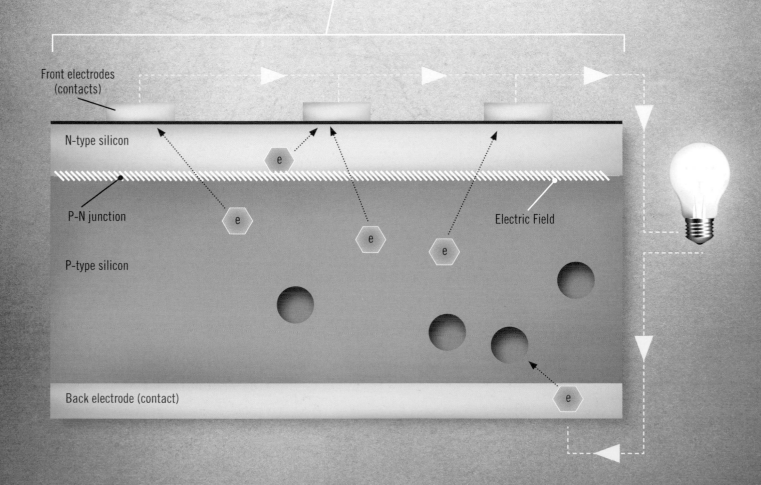

Front electrodes (contacts)

N-type silicon

P-N junction

Electric Field

P-type silicon

e

e

e

e

e

Back electrode (contact)

One of the Vanguard satellites at Cape Canaveral, Florida, in 1958

THE INVENTION OF PHOTOVOLTAICS

In 1839, Alexandre-Edmond Becquerel, a 19-year-old French prodigy, first demonstrated that light could create an electric current. Experimenting in his father's laboratory, Becquerel dipped an electrode into a beaker filled with a conductive solution and exposed the beaker to sunlight. As fast as you can say "Voilà!" Becquerel discovered that, indeed, sunlight had the ability to create electrical energy.

Then, in the 1870s, an English electrician named Willoughby Smith was using selenium to test for faults in underwater telegraph lines for the Gutta Percha Company. Smith noticed that the mineral conducted electricity very well in sunlight but less so in darkness. He decided to write an article on the subject creatively entitled "Effect of Light on Selenium During the Passage of an Electric Current," which the journal *Nature* published in its February 1873 issue. Two American scientists—William Adams and Richard Day—read Smith's article and started experimenting with light and selenium themselves. Soon, they discovered that sunlight generates a flow of electricity through the mineral. Eventually the two would author *A Substitute for Fuel in Tropical Countries*, the first book on solar energy. Adams and Day even invented a 2.5-horsepower steam engine driven entirely by sunlight.

Not long after Adams and Day's discovery, Charles Fritts, an American inventor, was experimenting with selenium in order to create more powerful electric currents. He put a layer of selenium on a metal plate coated with gold leaf and found that, when placed in sunlight, the combination produced much more electricity, though still not enough to be of much use. Still, Fritts had invented the first solar photovoltaic (PV) cell. At the time, however, few took much notice of his invention.

A backyard pool heated by solar panels in 1965

In 1905, Albert Einstein wrote about the mechanism of light exciting electrons (the photoelectric effect). But it took until 1921—when Einstein's insight won him the Nobel Prize in physics—for the idea of solar power to really take off. Different manufacturers began experimenting with using PV cells to power appliances. The first applications, however, weren't particularly practical. Though Bell Laboratories manufactured a PV cell in 1954, it was used mostly as an oddity, powering children's toys. The power created by the first solar cells was simply too expensive. In relative terms, electricity generated by these early PV cells cost over 100 times the price of producing the same amount of electricity using coal.

The design and materials of PV cells would improve with the advent of the space age. There was no practical way to launch a satellite powered by an internal coal-fired generator. And even with the best batteries, satellites would run out of energy too quickly to be of much use. The obvious solution was PV cells. By coating the outside of the satellite with solar panels, satellites could be powered almost indefinitely. So in 1958, the United States launched the Vanguard 1 satellite, which was covered in PV cells that powered the satellite until 1964, when its signal was lost. (The Vanguard 1 is the oldest human-made satellite still in orbit.)

In the late 1960s, energy giant Exxon gathered a team of researchers to look at projects that would provide the energy supplies of the future. The group assumed that electricity from conventional sources would be so expensive by the year 2000 that the world would be scrambling for cheaper alternatives. One scientist, Elliot Berman, joined the team in 1969 and suggested using thin layers of silicon as a conductive material to produce solar power at cheaper costs.

Through much trial and error, the group finally settled on a design that used thin wafers of silicon with electrodes printed directly onto one surface. By gluing an acrylic covering on the front and a crude circuit board on the back, the team designed a solar panel that was cheap enough to have a shot at competing against conventional sources of electricity. Exxon quietly started buying up scrap silicon from other industries and launched a subsidiary company called Solar Power Corporation (SPC). By 1973, using the design Berman's team had developed, SPC was churning out the first commercially available silicon-based solar panels that could produce electricity for less than $20 per watt, a five-fold decrease in cost over previous designs.

Even at this drastically reduced cost, however, SPC's solar panels could not compete with the $2 to $3 per watt cost of electricity generated from coal. But subtle design improvements made over 30 years, coupled with steadily increasing costs for conventional electricity, turned solar power into a competitive source of energy. By 2012, over 63 gigawatts of solar power had been installed worldwide, a ten-fold increase over the previous five years. According to the International Energy Agency (IEA), solar PV will represent over 13% of all the new electricity capacity installed by 2035.

CONCENTRATED SOLAR POWER

According to a famous legend, Archimedes, the Greek mathematician and inventor, used mirrors to concentrate sunlight on a fleet of Roman ships invading the city of Syracuse in 214 BCE. Before long, the ships burst into flame. Although there is much debate about whether Archimedes' famous "heat ray" actually worked, a 1973 experiment by a Greek scientist proved it was possible to burn a plywood ship covered in tar with the concentrated light of 70 copper-coated mirrors. In Archimedes' day, it was not uncommon for ships to be covered with a thin coating of tar to help keep them waterproof.

A 1642 engraving reimagining the mirror Archimedes used to focus sunlight on and burn the Roman fleet attacking Syracuse in 214 BCE

Whether the story about Archimedes is truth or myth, it is certainly true that sunlight can be focused to create enormous amounts of heat. Concentrated solar power (CSP) uses mirrors or lenses to concentrate a large amount of solar energy onto a small area. This concentration creates heat, which is then used to turn a turbine, just like in a conventional thermoelectric generator.

The most common CSP configuration uses parabolic troughs coated on the inside with reflective material. The troughs concentrate sunlight onto a tube positioned along a focal line a few feet away. Inside the tube, a special fluid (usually some form of molten salt) is continuously heated until it reaches temperatures up to 350 degrees Celsius (over 660 degrees Fahrenheit). The fluid flows through the tubes and is used to generate pressurized steam that is then forced through a turbine. As magnets on the turbine spin, they generate an electromagnetic current that is transmitted through wires as electricity.

Often called solar-thermal generation, commercially available CSP components were first manufactured in the United States in 1984. Since then, almost two gigawatts of CSP capacity has been installed worldwide. However, as cheaper designs and materials have become available, CSP is expected to become a growing source of electricity. One study by Greenpeace International and the European Solar Thermal Electricity Association (ESTELA) projects that, by 2050, more than 1,500 gigawatts of CSP will be installed, supplying as much as 25% of global electricity demand at that time.

A parabolic trough CSP system

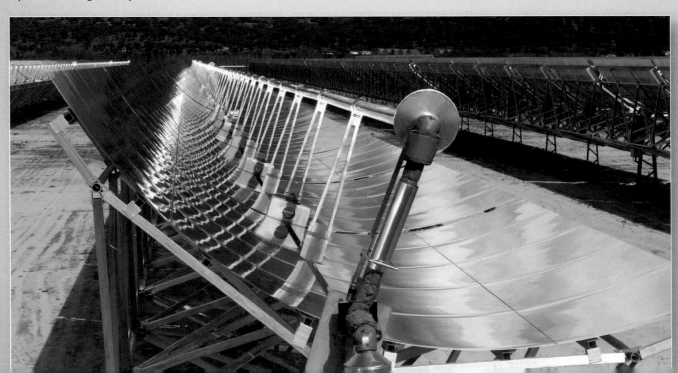

SOLAR POWER TOWER

Solar power towers, also known as heliostat power plants, are really a type of concentrated solar power. Rather than using mirrored troughs to focus sunlight on a liquid-filled tube close to the mirror, a power tower uses an array of flat mirrors called heliostats to focus sunlight on a central collection tower. A fluid (usually some form of molten salt) in the tower is used to collect the energy and generate steam with temperatures up to 500 degrees Celsius (over 930 degrees Fahrenheit). The superheated steam is used to drive turbines just like in any other CSP (or thermoelectric) plant.

In 2009, Abengoa Solar, a Spanish company, began operation of the world's largest solar power tower near Seville. The PS20 tower can generate up to 20 megawatts of electricity to power nearly 10,000 homes. In 2013, however, Google, NRG Energy, and BrightSource Energy are set to begin operating a massive solar tower complex in California's Mojave Desert that will generate up to 377 megawatts of electricity, enough to power 140,000 homes. The Ivanpah Solar Electric Generating System will use over 170,000 heliostats to concentrate sunlight onto a 450-foot central tower, the tallest solar power structure ever built.

The PS20 solar power tower near Seville, Spain

A solar furnace uses concentrated solar energy to produce superhigh temperatures for industrial uses, though usually not to produce electricity. The furnace in Odeillo, France, uses 63 mirrors (lower right) that track the sun and redirect sunlight toward a 2,000-square-meter (21,528-square-foot) parabolic reflector consisting of 9,500 curved panes of mirrored glass. The reflector concentrates the beams of light onto a focal point, producing temperatures of more than 3,000 degrees Celsius (5,432 degrees Fahrenheit).

POWER OVER POWER:
THE PERFECT SOLAR STORM

Coronal mass ejections have the potential to inflict massive damage on the modern electricity infrastructure. A large CME can contain billions of tons of highly charged solar plasma traveling at breakneck speed. As these charged particles slam into Earth's magnetosphere, they can create massive electric currents that follow Earth's horizontally oriented magnetic field lines all the way from the upper atmosphere into the surface of the planet. Electrical charges always want to move from areas of high voltage to areas of lower voltage. High-voltage transmission lines, therefore, act almost like antennas, attracting electromagnetic currents like a drain attracts water. They also tend to be north-south oriented and, therefore, contain multiple points along the lines where storm-induced currents can jump from the ground into the grid, frying wires and components as they rush toward points of lower voltage.

For two days in February 2010, senior government officials of the United States, Sweden, and the European Union, along with a handful of specially selected representatives from the private sector, gathered quietly in a conference room in Boulder, Colorado. They had been called together by the Federal Emergency Management Agency (FEMA) and the Department of Homeland Security (DHS) to run a simulation: what would happen if, right at that moment, the North American power grid were suddenly struck by a severe solar storm?

The results were frightening. According to the world's best experts, the storm would spark cascading power outages throughout the eastern and mid-Atlantic states and large parts of eastern Canada within the first hour. Power stations across the Northern Hemisphere would begin reporting transformer failures. Lacking backup transformers (and with virtually no North American transformer manufacturers), repairs and replacements would take several weeks to several months. The simulators estimated that emergency response personnel would face critical infrastructure failures within the first few days as water distribution, sewage, medical care, phone service, and fuel supply systems ground to a halt. Loss of cellular communications and Global Positioning System (GPS) would impair financial systems and severely delay response and recovery. Utility workers in affected areas would soon abandon their posts to provide for their families, as civil society collapsed around them.

It would be easy to dismiss this dire scenario as fiction, a kind of postapocalyptic fantasy that is more the stuff of zombie films than public policy. But the findings of this group of world experts were credible enough to compel Britain's chief science advisor to warn at the time that a severe solar storm would be a "global Katrina," lasting for several years and costing the world's economies as much as $2 trillion.

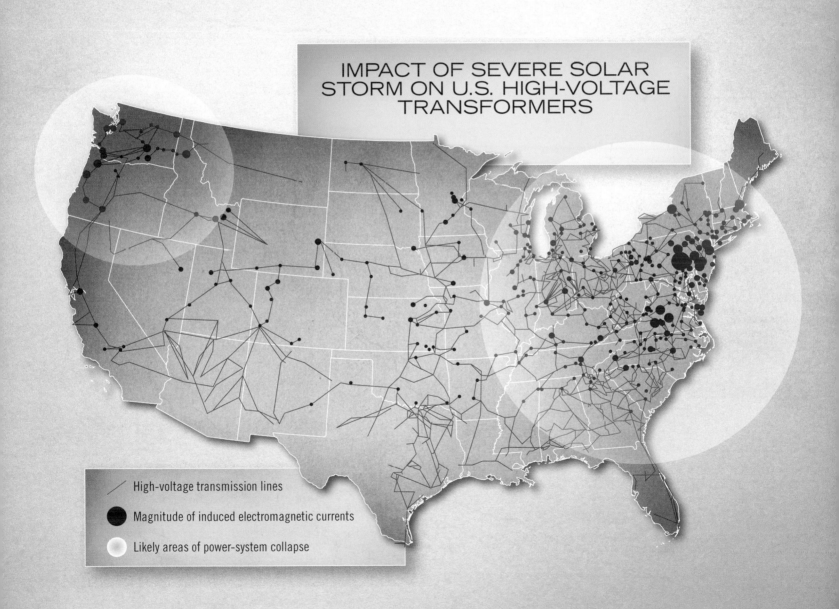

IMPACT OF SEVERE SOLAR STORM ON U.S. HIGH-VOLTAGE TRANSFORMERS

⁄ High-voltage transmission lines

● Magnitude of induced electromagnetic currents

○ Likely areas of power-system collapse

TRANSFORMER VULNERABILITY

The scariest part of FEMA's 2010 solar storm scenario was how a large coronal mass ejection would destroy critical high-voltage transformers. When they are saturated with current, transformer cores, which are essentially big magnets, heat up and the internal components literally melt. Extra-high voltage (EHV) transformers (the kinds used on long transmission lines delivering power from, say, Quebec to New York City), introduce an additional concern. Their design actually compounds the effects of storm-induced currents, causing them to become saturated with current more quickly than transformers used on low-voltage portions of the electricity grid.

An earlier solar storm simulation conducted in 2007 found that an especially strong CME would damage up to 350 EHV transformers beyond repair. At the time, those 350 transformers represented over 97% of all the EHV transformers in New Hampshire, 82% in New Jersey, 72% in Oregon, 40% in Washington, and between 24% and 75% of all EHV transformers in every state along the east coast of the United States, from South Carolina to Maine.

Nearly all transformers are manufactured outside North America. Even under normal conditions, replacing an EHV transformer can take up to 24 months and cost more than $10 million. Moreover, each transformer (even from the same manufacturer) can contain subtle design variations, complicating efforts to replace several at once. Loss of these transformers could cause wide-scale damage, collapsing power systems in highly populated areas of the mid-Atlantic and Pacific Northwest for several years.

An extra-high voltage transformer

THE CARRINGTON EVENT

In the nineteenth century, Richard Carrington was one of England's foremost solar astronomers. He charted sunspot activity by using his telescope to project an 11-inch-wide image of the Sun onto a screen and painstakingly drew what he observed. On the morning of September 1, 1859, Carrington was sketching an enormous group of sunspots when suddenly two brilliant beads of blinding white light appeared over where the spots had been. As he watched, the beads intensified and then gradually shrank to pinpoints until they finally disappeared.

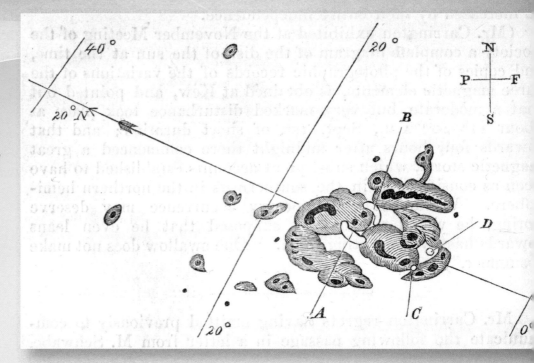

While observing sunspots, Richard Carrington witnessed "two patches of intensely bright and white light" (points A and B). He observed the light rapidly expand and then fade, noting that "The last traces were at C and D, the patches having travelled considerably from their first position and vanishing as two rapidly fading dots of white light."

Just before dawn the next day, the skies erupted in auroras so brilliant they could be seen as far south as El Salvador and the Bahamas. Though electrical transmission was not yet widespread, telegraph systems were—and they went haywire around the world, shocking startled telegraph operators and setting brush fires in some arid regions of the southwestern United States. Even after disconnecting the batteries used to power the lines, many operators could continue to transmit messages for several hours.

What Carrington had witnessed was the second largest coronal mass ejection ever recorded, a solar storm so powerful that New Yorkers could read their newspapers at midnight by the light of the aurora it created as charged particles from the Sun crashed into the Earth's magnetosphere. The storm was named the Carrington Event after the astronomer correctly deduced that it originated from the solar activity he had witnessed the day before. In 2008, the National Academy of Sciences predicted that if a storm the size of the Carrington Event struck today, it would cause up to $2 trillion in damages and would take 4 to 10 years to repair.

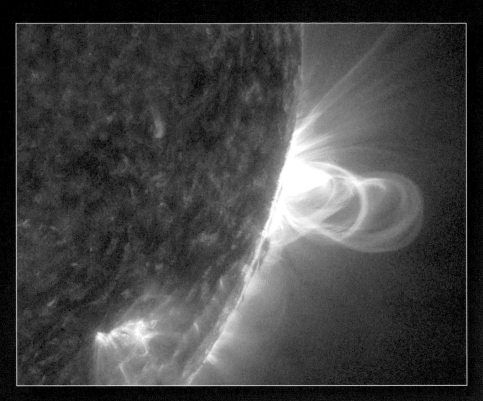

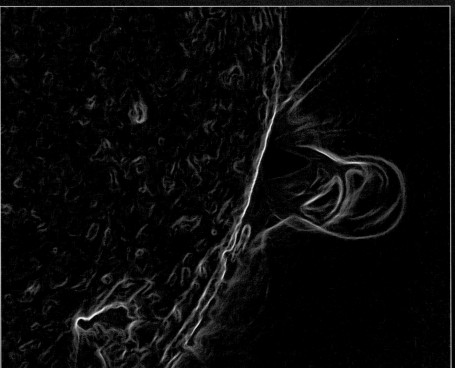

Scientists have long known that coronal mass ejections are associated with tightly wound magnetic flux ropes. What they didn't know was which came first, the CME or the flux ropes. On July 18, 2012, NASA's Solar Dynamics Observatory captured a small solar flare that produced magnetic flux ropes that twisted into figure eights. About eight hours later, another flare erupted from the same place. This time, the flux ropes snapped apart, sending tons of solar material into space in a classic CME—and providing solid evidence that twisted flux ropes precede CMEs.

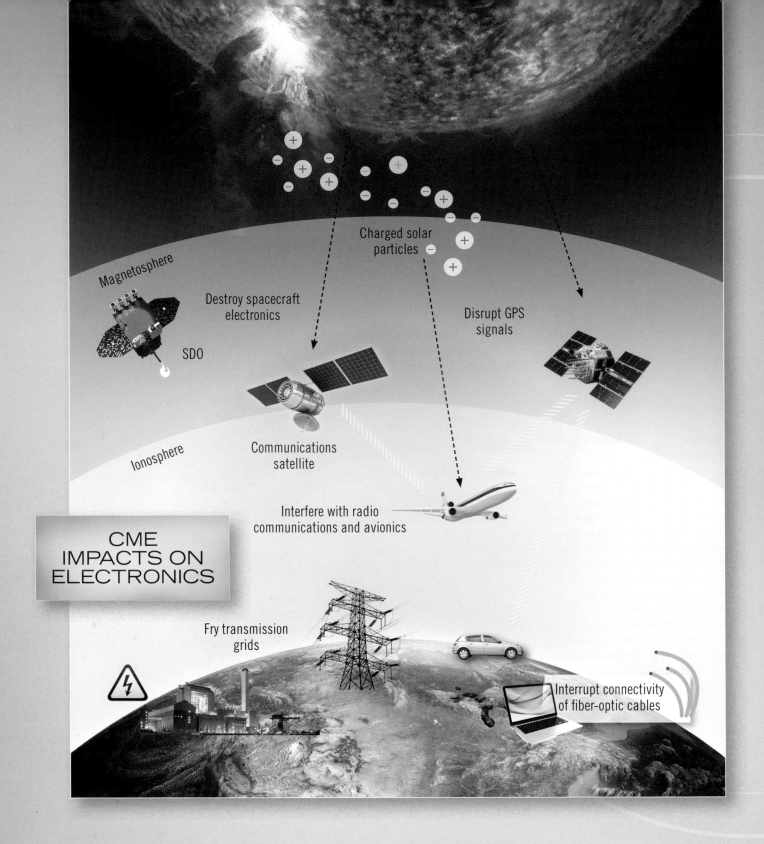

Magnetosphere

Destroy spacecraft
electronics

SDO

Charged solar
particles

Disrupt GPS
signals

Ionosphere

Communications
satellite

Interfere with radio
communications and avionics

CME
IMPACTS ON
ELECTRONICS

Fry transmission
grids

Interrupt connectivity
of fiber-optic cables

THE SUN AND MODERN ELECTRONICS

Almost all modern electronics would be vulnerable to very large coronal mass ejections. This is because many modern electronics depend on circuit boards or tiny computer chips with copper wiring that may conduct radiant electromagnetic energy during a solar storm. A particularly large CME could produce geomagnetic currents strong enough to burn circuitry with a jolt of electric current. Just about anything—from high-end dishwashers to electronic car ignitions—could be affected.

During a solar storm, it is not just electrical transmission lines that can be damaged. Coaxial cable for television and (increasingly) Internet connections may be vulnerable as well. Solar storms can create varying magnetic fields and induce voltage spikes in the center of the cable, causing an increase or decrease in voltage from the cable's power supply. These voltage spikes can overload the electricity on the cable system and knock out components connected to the line. Even fiber-optic cables may be at risk since many still require metallic (and conductive) wires that run along the length of the line and provide electricity to the amplifiers needed to boost signal strength.

Solar storms can also cause serious damage to Earth-orbiting satellites, especially those in geosynchronous orbits, which tend to be farther from the protection of Earth's magnetosphere. (A geosynchronous orbit is a high-altitude orbit that matches Earth's rotation, allowing satellites to remain directly over one spot on the planet.) Many communications satellites are put in geosynchronous orbit and can suffer damage during a solar storm. High-energy particles can damage electrical components on or within the satellites. But even satellites with some form of protection can be damaged when charged particles bombard the shielding, building up an electrical charge that eventually discharges into the internal components.

Disruption of satellite services can affect major news wire-service feeds, network television, weather data, cell-phone service, ATMs, airline tracking systems, and critical military communications. Satellites in the Global Positioning System (GPS) may be particularly at risk. Magnetic currents produced during a solar storm can generate field distortions and produce erroneous compass readings. Many GPS satellites depend on attitude control systems that make fine adjustments to orient the satellites based on the position of Earth's magnetic poles.

These dire consequences are precisely why many governments have made monitoring and forecasting solar activity new national priorities. At the U.S. National Oceanic and Atmospheric Administration's Space Weather Prediction Center, a team of space weather forecasters continuously monitor data from ground- and space-based sensors looking for anomalies that might indicate a large CME was on its way toward Earth. With enough warning time, they can alert vulnerable industries so satellite operations can be adjusted and power grids reconfigured to minimize disruption.

Individuals can minimize the impact of severe solar storms by making a few preparations. If possible, have a source of off-grid backup power (a gas generator, solar panels, or a small-scale wind turbine) to provide electricity should central electrical service be impacted. Have an emergency pack ready, with flashlights, batteries, nonperishable food, and bottled water as well as a short-term supply of cash (since credit card and

Although not caused by a CME, a widespread blackout in August 2003 gave a preview of what might happen after a severe solar storm. In New York City, mass transportation came to a standstill.

ATM systems could be out for an extended period). Connect your electronic devices to an uninterruptible power supply (UPS), which looks like a standard surge protector but is equipped with batteries that keep computers, cell phones, and other electronics running smoothly during major power fluctuations. This will ensure that, even if you cannot use the electronics until power and cable services are restored, they will not be damaged during initial CME impacts.

THE QUEBEC BLACKOUT

Since the advent of the modern electricity grid, the most famous solar storm was a geomagnetic storm that struck Canada in March 1989. At 2:44 a.m. on March 13, 1989, a large energy jolt in the Earth's magnetic field pulsed along the U.S.–Canadian border. Currents induced by a coronal mass ejection had created huge differences in voltages at different points in the high-voltage transmission network spanning the border. Because voltage asymmetries reached 15%, equipment designed to protect the grid from voltage spikes began to shut down parts of the system. Voltage dropped so precipitously between five high-voltage lines in the La Grande transmission network that it separated from the Hydro-Quebec grid to avoid being fried.

Almost immediately, over nine gigawatts of electricity being pumped into the Quebec grid evaporated. Protective controls near large cities were designed to handle electrical spikes, but not to recover from such an enormous loss of electricity. Within 25 seconds, as various portions of the grid shut down to protect themselves from huge voltage fluctuations, the entire network collapsed, leaving millions of people without electricity for the next nine hours.

The National Oceanic and Atmospheric Administration (NOAA) ranked the 1989 storm the third largest recorded since rankings began in 1932. Until recently, many experts in the electricity sector believed solar storms of that size represented the worst-case scenario for the North American bulk power grid. But an analysis of historical data has revealed that the northern latitudes have had near misses from storms roughly four times larger on at least three occasions since 1972.

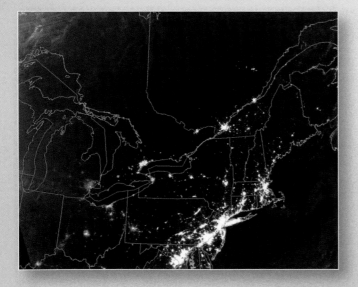 

If there were before-and-after satellite photographs of the 1989 Quebec blackout, they may have looked something like these. The photo on the left was taken about 20 hours before a massive blackout on August 14, 2003. The photo on the right was taken several hours later, when much of southeastern Canada and the northeastern United States were still without power.

TRANSPOLAR FLIGHT

During the Cold War, the Arctic zone provided a military buffer between the United States and the Soviet Union. As a result, there were few civilian transpolar flights. When hostilities diminished with the breakup of the Soviet Union in the late 1980s, however, both sides became less concerned about a transpolar attack. Airlines realized that flying over the poles rather than along the lower latitudes could dramatically reduce flight distances and save fuel, especially for routes between the United States and Asian cities like Beijing, Hong Kong, and Singapore. For example, Emirates Airlines estimates that it saves about 2,000 gallons of fuel by routing its Dubai to San Francisco flight over the North Pole. And it is not alone.

By 2008, United Airlines scheduled over 1,400 trans-Arctic flights, including its daily runs between Chicago and Hong Kong and Chicago and Beijing. Other U.S. airlines fly daily routes over the Arctic, including Northwest Airlines, which has four transpolar flights weekly between Detroit and Beijing. Only one airline—Qantas—flies a route over the South Pole, making five weekly flights between Sydney and Buenos Aires.

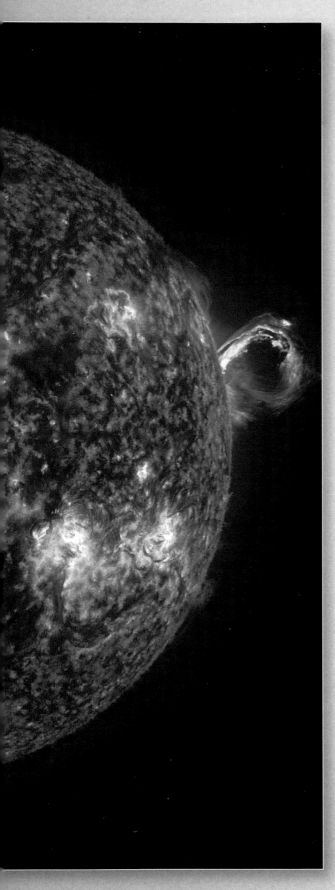

A beautiful view of a coronal mass ejection on May 1, 2013, as it lifted off the surface of the Sun before blasting off into space.

Somewhere above 82 degrees latitude north (or below 82 degrees south), most transpolar flights lose line-of-sight with communications satellites and have to switch to high-frequency (HF) radio communications. Some solar storms can cause disturbances in the ionosphere that disrupt HF radio signals, endangering the airplanes. For example, charged particles from the Sun may penetrate the ionosphere and absorb radio frequencies, causing communications to fade out. Some solar flares (particularly those within the frequency of X-rays) can also cause HF radio fadeouts on the sunward side of the Earth.

Because of their reliance on HF radio communications, airlines alerted to impending solar storms must divert transpolar flights to conventional routes. During a particularly strong solar storm in January 2012, Delta, Qantas, and Air Canada had to reroute transpolar flights between the United States and Asia. Fortunately, the airlines knew a solar storm was on its way. Without warning from the NOAA's Space Weather Prediction Center, these transpolar flights could have unexpectedly experienced a loss in radio communications and severe (and dangerous) navigational system errors.

THE SOLAR CYCLE

In 1826, the German pharmacist-turned-astronomer Samuel Heinrich Schwabe had an odd theory that a planet, which he tentatively named Vulcan, existed closer to the Sun than Mercury. Schwabe thought that the best way to discover this planet would be to look for its shadow as it passed between the Earth and the Sun. Fortunately for us, Schwabe's kooky theory motivated him to make precise observations of sunspots nearly every day for 17 years. Schwabe never found Vulcan, but he did notice that the number of sunspots seemed to vary over time and published a paper to that effect in 1843.

Schwabe's paper attracted the notice of Swiss astronomer and mathematician Rudolf Wolf. Wolf was so determined to verify period variations in sunspots that he gathered all of the data collected on them since astronomers had begun to observe them with telescopes in the seventeenth century. But to define sunspot activity over time, Wolf needed to devise a standard unit. So he created a formula that accounted for the number of sunspots, the number of sunspot groups, and minor variations in observed activity because of location or instrumentation used to record the activity. The result was what is known as the Wolf number, a standard measure of sunspot activity still used today.

After reconstructing all of the data he could find on sunspots as far back as 1610, Wolf used his calculation to determine that sunspots followed a cycle of about 5.5 years of increasing activity followed by about 5.5 years of decreasing activity, completing one solar cycle every 11.1 years. Recent research has revealed a more precise calculation of 10.66 years, but most people refer simply to an 11-year solar cycle. The period of increasing solar activity is known as solar maximum, while the period of decreasing solar activity is known as solar minimum.

Once scientists could identify the periodic cycle of solar activity, they were able to notice patterns in solar cycles. For example, evaluation of the available data revealed that particularly strong solar maximums tended to take less time to peak than weaker solar maximums, a phenomenon known as the Waldmeier effect. Recently, many astronomers have questioned whether the Waldmeier effect is real or a by-product of how Wolf defined a sunspot. Some scientists also suggest that the strength of solar cycles themselves wax and wane over a period of about 90 years, known as the Gleissberg cycle.

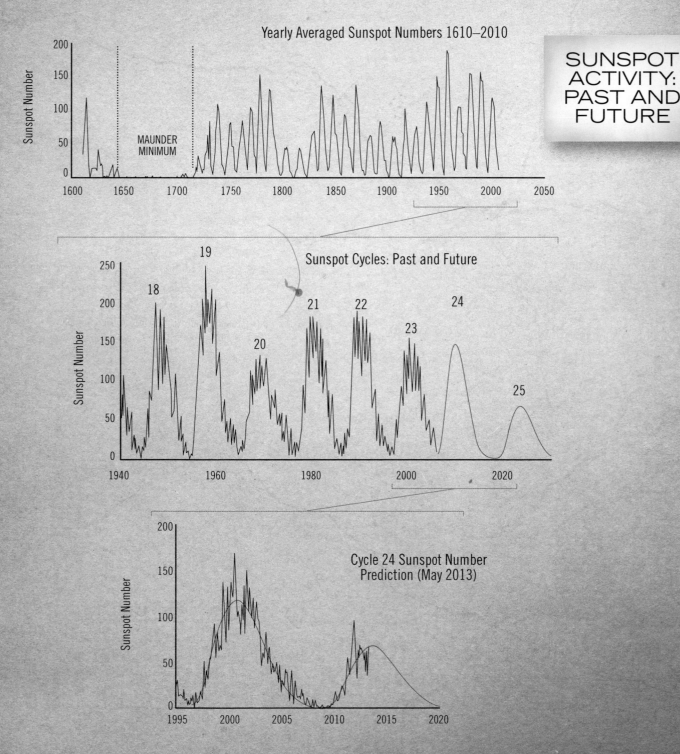

SUNSPOT ACTIVITY: PAST AND FUTURE

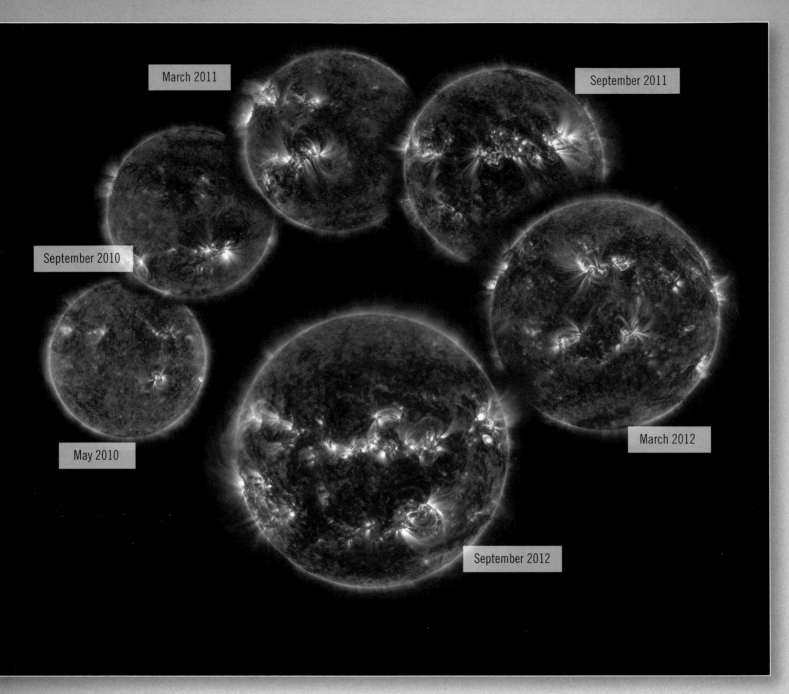

Starting in May 2010, six images from NASA's Solar Dynamics Observatory track the rising level of solar activity as the Sun progressed toward solar max in 2013.

Knowing that solar activity follows a predictable cycle is important because of the many ways an active Sun impacts our planet. During solar maximums, the Sun emits larger amounts of radiation in the extreme UV spectrum that can cause significant changes in the conductivity of the Earth's ionosphere. Solar maximums are also associated with stronger, more frequent solar flares and coronal mass ejections, which can cause massive disturbances in the Earth's geomagnetic fields.

Surprisingly, humans are exposed to more UVB radiation during solar minimums than solar maximums because of the impact of UV radiation on the Earth's protective ozone layer. Ozone forms in the upper atmosphere when UV radiation splits oxygen (O_2) into free-floating ions of oxygen. These highly reactive ions combine with regular oxygen molecules to form ozone (O_3). When solar radiation decreases during a solar minimum, fewer oxygen molecules are split and the concentration of protective ozone in the atmosphere decreases. As the ozone layer thins, more of the Sun's UVB radiation can penetrate to the Earth's surface.

THE LITTLE ICE AGE

Around the turn of the twentieth century, a British astronomer and mathematician named Edward Maunder and his second wife, Annie (a brilliant mathematician in her own right), photographed and studied sunspots for the Royal Observatory. While reviewing historical sunspot data, the couple identified an extended period between 1645 and 1715 when sunspots were unusually rare. During one 30-year timespan, astronomers had counted about 50 sunspots rather than the typical 40,000 to 50,000. In fact, in 1670 they observed no sunspots at all.

Many of Maunder's contemporaries blamed the lack of sunspot activity on shoddy observations. But the Italian astronomer Giovanni Cassini made painstaking observations of sunspots over much of the period. Polish astronomer Johannes Hevelius also published descriptions of the solar surface that he and his wife had made continuously between 1652 and 1685. The French astronomer Jean Picard (no relation to the fictional starship captain) made systematic observations of the Sun on every clear day from 1653 until he died in 1685. So the problem was not a lack of accurate data. Something had caused the Sun to go silent.

This prolonged period of minimal sunspot activity has become known as the Maunder Minimum and corresponds with the coldest part of the Little Ice Age, a period between about 1550 and 1850 when temperatures in Europe and North America plummeted. Winters were so bitterly cold that many farms and villages in Switzerland were destroyed by expanding glaciers. In 1658, the largest strait (known as the Great Belt) between the main Danish islands froze over, allowing the Swedish army to march across the sea and attack Copenhagen. The period also saw famines in Estonia, Finland, France, Norway, and Sweden. Agricultural output dropped so precipitously that one historian described Alpine villagers stretching their diminished stores of barley and oats by mixing ground nutshells into their bread flour.

In 1683, Londoners made the best of the unusually cold weather and held a Frost Fair on the Thames River.

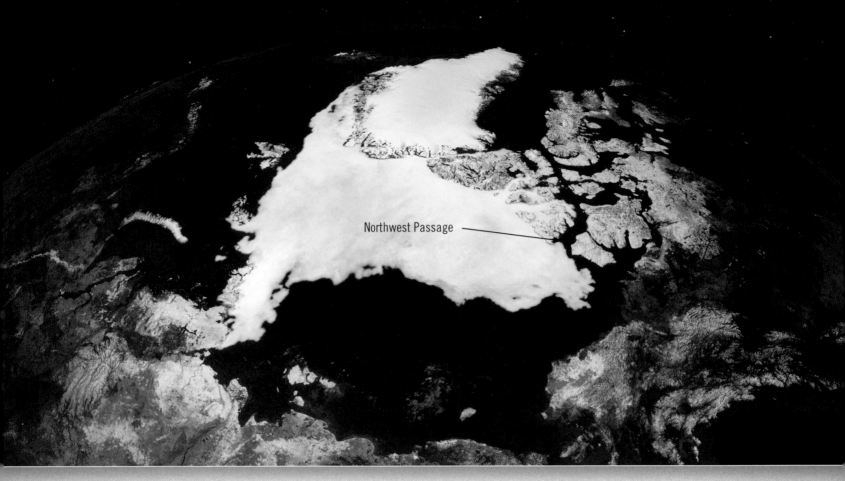

Northwest Passage

This data visualization from the Advanced Microwave Scanning Radiometer on NASA's Aqua satellite shows the Arctic sea ice on September 14, 2007, when the Northwest Passage was free of ice for the first time since satellite records began.

THE SUN AND CLIMATE CHANGE

In the 1930s, Charles Greeley Abbot, an American astrophysicist (and inventor of several solar-powered appliances, including the solar cooker and solar still), posited a connection between solar cycles and the Earth's climate. Although Abbot earned enough respect among the scientific community to be named secretary of the Smithsonian Institution, he was widely criticized for his theories. Unfortunately for Abbot, accurate measurements of solar radiation simply were not available in his day. Since the advent of satellites, however, scientists have taken more precise measurements that tend to support Abbot. As a result, a growing consensus within the scientific community is acknowledging the significant contributions solar activity makes to global climate change.

Though the correlation between the Maunder Minimum and the coldest period of the Little Ice Age has been noted for years, only recently have scientists been able to make a causal connection between especially weak solar minimums and extended periods of cold winters on Earth. In 2011, using data from NASA's Solar Radiation and Climate Experiment (SORCE), a team of British scientists uncovered a link between the variability of solar activity and the Earth's climate. The result has been a veritable war of words between climate scientists, most of whom believe that human activity is causing significant global climate change, and climate skeptics, who point to things like the variability of solar activity as the main cause.

The Intergovernmental Panel on Climate Change (IPCC) is a scientific body established by the United Nations in 1988 to investigate the causes and impacts of global climate change. Thousands of scientists, engineers, meteorologists, and other experts representing more than 120 governments voluntarily assess available data and write and review reports. In 1990, the IPCC completed its First Assessment Report, which concluded that emissions of certain greenhouse gases from human activity were contributing to a measurable increase in global temperatures. Since then, the IPCC has released three additional reports and one supplemental report, all confirming a link between human activities and observed global climate changes.

In early 2013, Alec Rawls, an American blogger who volunteered to review the IPCC's latest report (anyone may apply to review IPCC reports), leaked a draft of the report online. Rawls highlighted one paragraph in chapter seven examining the link between solar activity and the Earth's climate. In it, the drafters acknowledge that solar radiation appears to account for some of the observed warming of the Earth (though scientists are unsure of the exact mechanism). Rawls and other climate skeptics were quick to jump on the paragraph as evidence refuting the IPCC's claim that human activity is driving global climate change.

The truth is always a little more complicated. While variations in solar activity can (and do) impact Earth's climate, the current rate of global warming is almost certainly not the result of solar activity. Bob Berman, one of the world's foremost experts on all things solar, has noted that the Sun is only one of four factors that substantially affect the Earth's climate: volcanic dust, cyclical weather events like El Niño, greenhouse gas emissions, and variations in solar radiation. According to NASA research, sunspot activity can transfer to the Earth energy equivalent to roughly 15 years of greenhouse-gas emissions (at the rate humans were emitting greenhouse gases in 2010). While that amount is substantial, Berman claims that it has been dwarfed since 1994 by nonsolar factors, chief among them greenhouse-gas emissions.

To back up his claim, Berman notes that global temperatures peaked in the 1940s but then began falling for 30 years, even as carbon dioxide (CO_2) emissions increased. Then a strange thing happened. Global temperatures started climbing again, faster than ever recorded. At the same time, solar activity decreased. The Sun's brightness fell during solar cycle #20 (around 1970) and dropped even more during cycle #23 (around the year 2000). Yet global average temperatures continued to soar. In other words, according to Berman, the Sun *does* contribute to climate variability. But until around 1990, solar activity was merely masking the greater contribution being made by greenhouse-gas emissions. In fact, by decreasing its activity, the Sun has partially counteracted human contributions to global warming. But if solar activity starts to increase again, things may get a lot hotter.

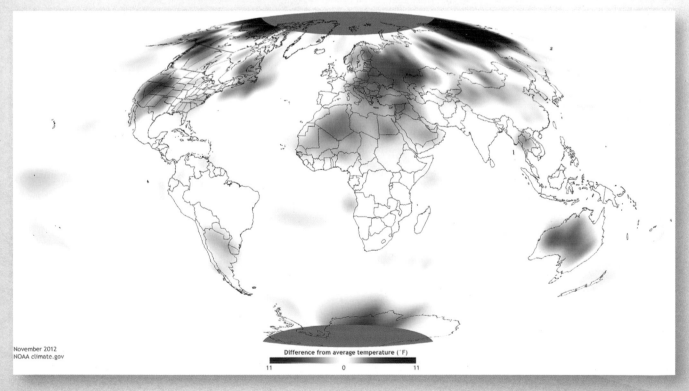

According to NOAA's National Climatic Data Center, November 2012 was the fifth warmest November since record-keeping began in 1880. On the map, areas that experienced warmer temperatures than the 1981–2010 average are in red and cooler temperatures are in blue.

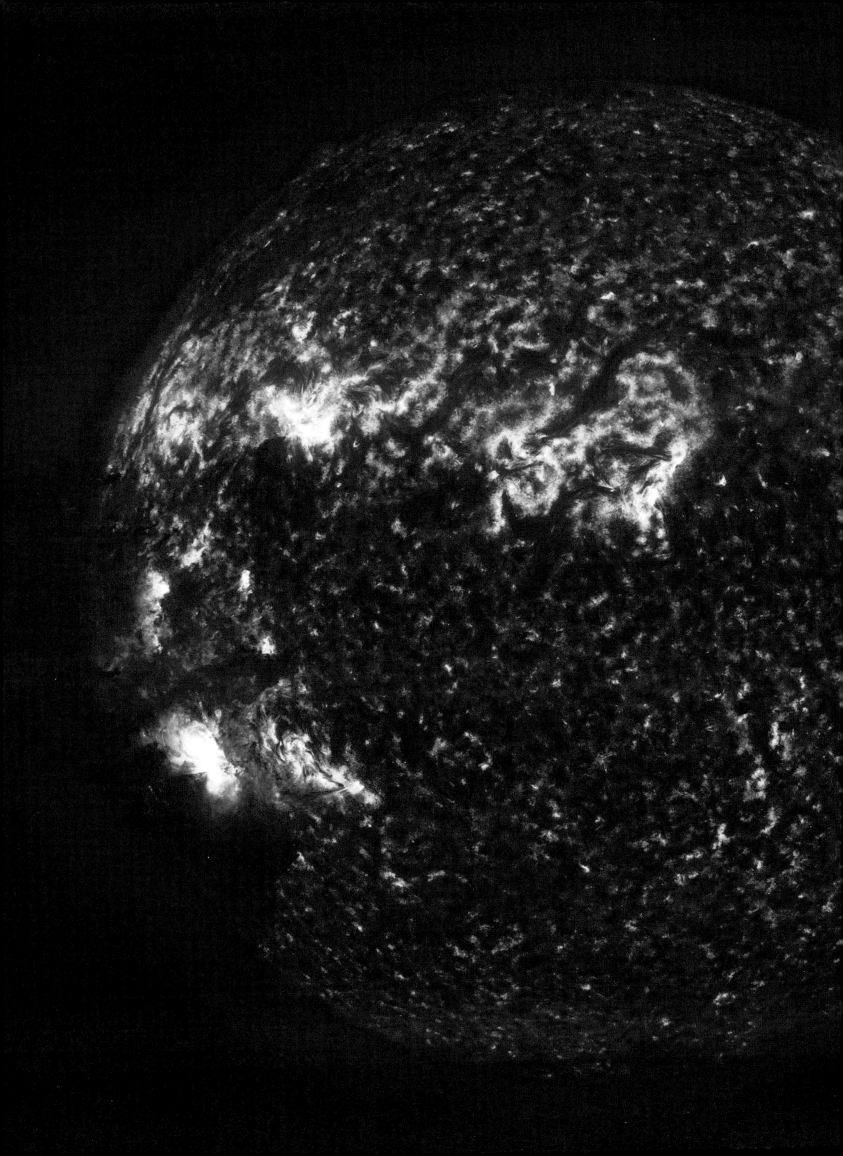

7

THE FUTURE OF OUR SUN

In the prime of our Sun's life, it is continuously converting hydrogen to helium through the process of nuclear fusion. Over the course of time, however, this constant fusion changes the Sun's interior. At its center, where high temperatures and pressures accelerate the fusion process, the amount of helium increases compared to the core's outer edges, where fusion is slightly slower. As more and more of the inner core is converted to helium, the nuclear fusion process will move outward, searching for more hydrogen to fuse. Meanwhile, the helium-rich inner core will keep growing.

RUNNING OUT OF GAS

Within the Sun's core there are always two forces at play—the outward-pushing pressure of nuclear fusion and the inward-pulling tug of gravity. As the fusion process moves away from the core, the outward pressure will weaken, but the inward pull of gravity will remain the same. Once a substantial amount of the hydrogen fuel in the Sun's core is converted to helium (when the Sun is approximately 10 billion years old, about 5.4 billion years from now), gravity will start to overcome fusion and the helium core will begin to contract.

This contraction will start to break the atomic structure of particles deep in the Sun's core, releasing energy that will be transferred outward, driving up temperatures on the outer edges of the core, where hydrogen fusion continues. The extra heating of the outer core will cause the Sun's remaining hydrogen to fuse more quickly. Within a relatively short period of time, the Sun's structure will change from a furiously burning inner core surrounded by a thick layer of hydrogen to a furiously burning thin shell of hydrogen surrounding an inner core of inert helium.

MISSING MASS?

For an easy way to calculate the Sun's loss of mass during hydrogen fusion, look at a periodic table, which shows one hydrogen atom as having a mass of 1.008 daltons (a dalton is the standardized unit used for indicating mass on the atomic level). One atom of helium, according to the table, has a mass of 4.0026 daltons. During fusion, the Sun converts four atoms of hydrogen into one atom of helium. But 4 hydrogen atoms at 1.008 dalton per atom = 4.032 daltons, just a tad more than the 4.0026 daltons of a helium atom (specifically, 0.7% more). The missing mass is what has been converted to energy. Although it does not seem like much, that 0.7% makes up the bulk of energy our Sun will emit over the course of its life and is far greater than any energy source on Earth.

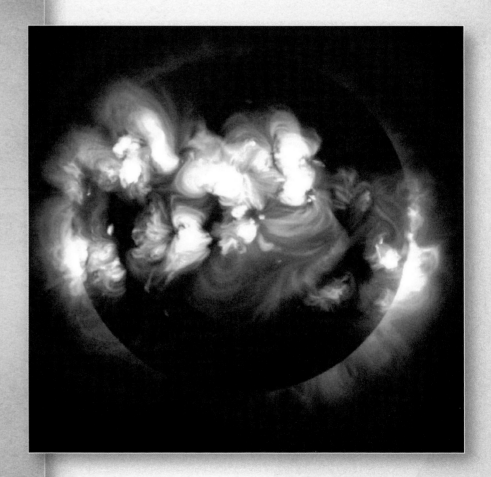

HYDROGEN DEPLETION IN THE SUN

Oddly enough, as the Sun is running out of hydrogen fuel, it will actually burn brighter. The extra energy released from atoms being crushed in the inner core will cause the fusion process in the hydrogen shell to quicken and exert pressure on the Sun's remaining, non-burning outer layers. Eventually, this outward pressure will become so great it will overcome the Sun's inward gravitational pull. Even while the fiery hydrogen shell around the Sun's outer core is heating up, the outward pressure will cause the rest of the outer layers to begin rapidly expanding and cooling. This process of inward heating causing outward expansion and cooling will continue as the Sun swells in size.

PRIME OF LIFE

HYDROGEN
CORE FUSION

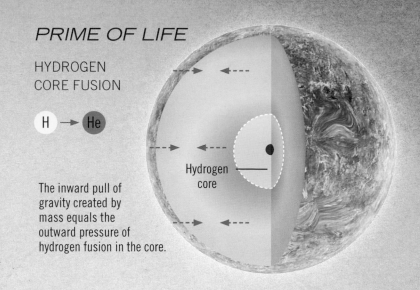

H → He

The inward pull of gravity created by mass equals the outward pressure of hydrogen fusion in the core.

Hydrogen core

HYDROGEN DEPLETION

HYDROGEN
SHELL FUSION

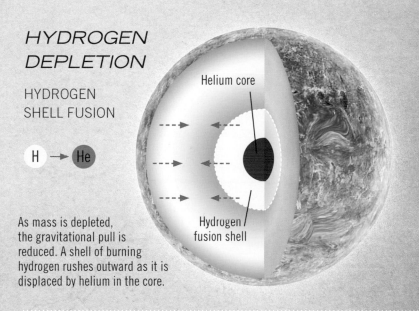

H → He

As mass is depleted, the gravitational pull is reduced. A shell of burning hydrogen rushes outward as it is displaced by helium in the core.

Helium core

Hydrogen fusion shell

RED GIANT PHASE

HELIUM
CORE FUSION

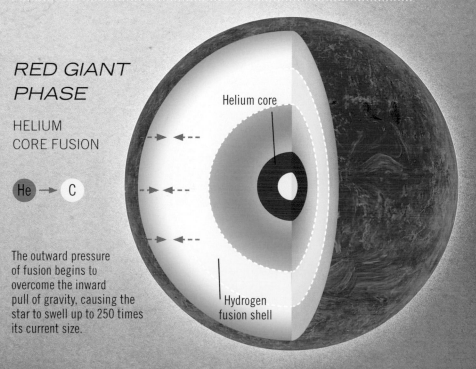

He → C

The outward pressure of fusion begins to overcome the inward pull of gravity, causing the star to swell up to 250 times its current size.

Helium core

Hydrogen fusion shell

THE AGE OF OUR SUN

Nebula

> 4.6 billion
years ago

Protostar

4.6 billion
years ago

Birth

**(Hydrogen burning
begins)**

4.5 billion
years ago
--
Sun's age:
0

G-Type Main Sequence

4.5 billion
years ago to 5.4 billion
years from now
--
Sun's age:
Birth to
10 billion years

Hydrogen Exhaustion

5.4 billion
years from now
--
Sun's age:
≈10 billion years

Red Giant

5.4 billion
years from now
--
Sun's age:
10 to 11 billion years

White Dwarf

6.4 billion
years from now
--
Sun's age:
11 billion to ≈10 quadrillion years

Black Dwarf

≈10 quadrillion
years from now
--
Sun's age:
>10 quadrillion years,
in theory

THE RED GIANT PHASE
(10 BILLION YEARS TO 11 BILLION YEARS)

As the Sun's outer layers cool to a little over 4,700 degrees Celsius (almost 8,500 degrees Fahrenheit) about 5.4 billion years from now, it will enter the red giant phase of its life. Its appearance will change from brilliant yellow to luminous red. Though it will change color, it will actually appear several hundred times brighter because of its enormous size. In fact, during its red giant phase, the Sun's radius will expand by a factor of at least 200, until its surface reaches about as far as Mercury's present orbit.

At the same time, gravity at its core will compress about 25% of the Sun's total mass into a ball about 1/1000th its current size (a little larger than the Earth). The density of the core will increase from 150,000 to almost 1 billion kilograms per cubic meter (9,364 to over 62.4 million pounds per square foot). But its cooler outer layers will become increasingly thin. Outward pressure will blow more and more matter into space, until, over a period of about 1 billion years, the photosphere becomes ill-defined and the layers outside the core transition into a corona of enormous size.

THE BIG DRIFT:
THE SUN'S REDUCED GRAVITY

If the planets remained in their current orbits, fairly simple calculations indicate that, during its red giant phase, the Sun would expand until its equator stretched well past Mars. Under this scenario, all the inner planets—Mercury, Venus, Earth, and Mars—would be swallowed up. But as the Sun's mass is turned from hydrogen into helium, part of that mass is converted to energy. The Sun's mass is also the source of its gravity, so as the Sun loses mass over its lifetime, its gravitational pull will be reduced. The reduction in the Sun's gravity will, of course, impact the orbit of planets in our solar system. In fact, an orbit is simply the gravitationally curved path of an object around a point in space. In the case of the Earth, that point is near the center of the Sun.

Interestingly, the Sun's loss of mass will not reduce its gravitational pull on Earth at a constant rate. This is because the Earth (like all the planets in our solar system) exerts a pull of its own, partly because of its mass and partly because it is spinning around the Sun, converting kinetic energy into a type of centripetal force. Astronomers calculated that, during initial changes in the Sun's mass, the Earth's orbit would change in proportion. But as the Sun's mass loss approaches 50%, the gravity of Mars and the outer planets would start to overcome the Sun's gravitational pull. As a result, their orbits should expand rapidly. At 50% mass loss, the outward pull of these planets should win out entirely, whipping them off into interstellar space.

THE SUN'S OUTER SWELLING

During its red giant phase, about 5.4 billion years from now, the Sun will swell to about where the Earth currently orbits.

Mercury　　Venus　　Earth

WHEN THE SUN SWALLOWS THE EARTH

In 2008, two astrophysicists, Klaus-Peter Schroder and Robert C. Smith, created a model of the Sun's evolution in order to calculate how the Sun's change in mass will impact the orbit of the planets. According to the model, the Sun will begin to lose mass quickly when it enters its red giant phase. By the time it has swelled to its greatest size, it will have lost about 67% of its current mass. While the Earth's orbit may expand up to 50%, the model predicts that it won't happen fast enough to escape the Sun's expansion. Before the Earth can get far enough away, the expanding Sun will catch up to it. Schroder and Smith estimate that this will happen around 500,000 years before the Sun reaches its peak size. If their calculations are correct, had the Earth's original orbit been just 0.15 AUs farther out, the planet would have (barely) escaped being engulfed by the Sun.

Once inside the Sun's corona, the Earth will be bombarded by solar matter, its orbit will decay, and it will quickly spin into the inferno. Remarkably, except for the three inner planets, much of the solar system will survive, and even thrive . . . for a time. According to Schroder and Smith's model, the circumstellar habitable zone will have shifted to the outer solar system, allowing liquid water to exist well past the current orbit of Pluto.

BEYOND:
HUMANITY'S NEED TO ESCAPE

Well before the Sun engulfs the Earth during its red giant phase, solar radiation will have destroyed all life on the planet. Just as the Sun nears 10 billion years old, its energy output will be more than any remaining life on Earth (assuming there is any) could handle. The Sun will celebrate its 10-billionth birthday by heating the Earth to over 1,300 degrees Celsius (about 2,400 degrees Fahrenheit) and turning the Earth's oceans to vapor. By its 13-billionth birthday, the Sun will have melted the surface of the planet into a vast sea of lava. By the Sun's 14-billionth birthday, not a whisper of the Earth's atmosphere will remain.

On his *70th* birthday, the famed physicist Stephen Hawking gave a radio interview during which he advocated space colonization and warned that it is dangerous for the human race to put all of its eggs in one basket (or one planet). Assuming nothing else destroys us first, humanity has a while to plan our escape from the Sun's impending doom. Nevertheless, if there is any hope for our survival, we will have to find a way off the planet and into a permanent, habitable new home somewhere beyond this solar system.

In 2011, a team of researchers with the Kepler Space Observatory Mission identified 54 exoplanets—planets outside our solar system that orbit within the habitable zone of their stars. Of these, five were believed to be smaller than twice the size of Earth and, therefore, might have gravitational forces humans could withstand. Extrapolating from these results, the team estimated that there must be at least 500 million planets—among the 50 billion in our galaxy—that are within the habitable zone.

While 500 million planets sounds like a lot, Kepler scientists soon discovered several reasons that a suitable new home for humanity may be harder to find than they initially assumed. For one thing, 70% to 90% of all the stars in our galaxy are small red dwarfs that emit so little energy (relative to our Sun) that any Earth-like planet within the habitable zone would have to orbit very close to its host star. But at close distances, the red dwarf's tidal forces would be so strong that they would prevent the planet from spinning. One side of the planet would always face the star, while the other side would always face away. This would mean not only discomfort for a species accustomed to the Earth's cycle of day and night, but also that one side of the planet would be boiling hot while the other side would be freezing. Also, one side would constantly be exposed to radiation from the star while the other would be incapable of photosynthesis (as we know it). More importantly, a planet that cannot spin probably cannot produce a magnetic field strong enough to protect humans from the powerful solar flares most red dwarfs emit.

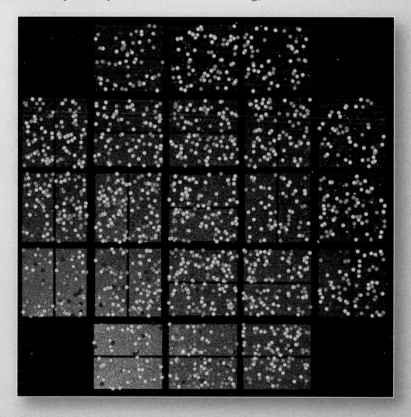

As of January 7, 2013, Kepler had identified 1,573 planets that appear to be within the circumstellar habitable zone of their stars, and more are being found all the time. Earth-size stars appear in blue, super-Earth-size stars in green, Neptune-size stars in orange, and giant-planet-size stars in red.

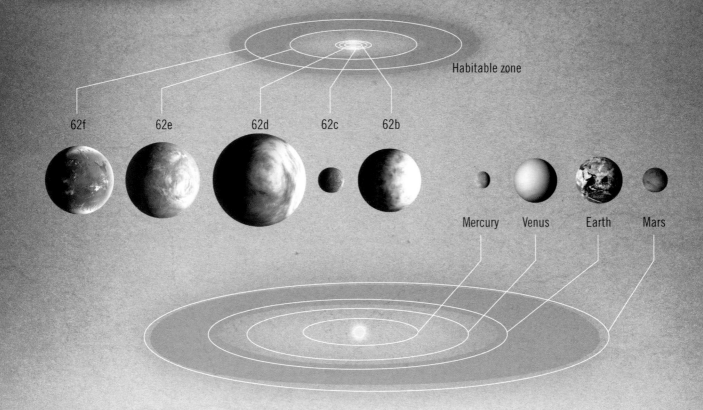

KEPLER-62
SYSTEM

62f 62e 62d 62c 62b

Habitable zone

Mercury Venus Earth Mars

The five-planet Kepler-62 system is compared to the inner planets of our solar system in this diagram. Kepler-62e is one of the highest-ranking exoplanets on the Earth Similarity Index. It orbits a dwarf star that is smaller, dimmer, and older than our Sun.

As of April 24, 2013, the Kepler team listed nine potentially habitable planets in its Habitable Exoplanet Catalog. Of these, Gliese 581g tied with newly listed Kepler-62e (discovered in April 2013) for the highest score on the Earth Similarity Index (ESI), a measure of the similarity of a planet to Earth on a scale from 0 (least similar) to 1 (exact replica). Gliese 581g—a planet thought to orbit the red dwarf Gliese 581, some 22 light years from Earth in the constellation Libra—achieved a score of 0.82 on the ESI, as did Kepler-62e.

Even with its high ESI score, Gliese 581g is a good reminder that a new Earth will be difficult to find. Gliese 581g suffers from all the problems that come with orbiting a red dwarf: it does not spin, a large portion of any water on the planet is likely frozen, and its surface temperature is estimated to range from –35 to –12 degrees Celsius (–35 to 10 degrees Fahrenheit).

Moreover, there is some debate over whether the planet exists at all. In 179 measurements of the Gliese star system taken over 6.5 years by the High Accuracy Radial Velocity Planet Searcher (HARPS), a high-precision spectrograph designed to find habitable exoplanets by spectroscopy, Gliese 581g never showed up. The Kepler team claimed that the researchers analyzing HARPS data made a methodological error that caused them to miss the planet—a claim the HARPS researchers deny. The debate likely will remain unresolved until better data is available. Nevertheless, Kepler's experience with Gliese 581g, initially touted as the most Earth-like planet yet discovered, is a sobering reminder that Earth is a pretty special place and replacing it could require sifting through millions of false hopes.

Assuming that our new home is one of the habitable exoplanets, humans will need to overcome the many difficulties involved in traversing the vast distances between stars. Interstellar travel requires developing a propulsion mechanism strong enough to propel a spacecraft (large enough to hold a sizeable population) to unbelievable speeds. According to Albert Einstein's theory of special relativity (think $E = mc^2$), mass and energy are different forms of the same thing. One unfortunate impact of this equivalence is the relativistic effect of velocity. That is, as a spacecraft reaches speeds approaching the speed of light, it gains relative mass, making it increasingly difficult to accelerate further. As a result of this effect,

Highly ranked on the Earth
the foreground of this artis
star and three other inner

any interstellar propu   hing other solar systems within a human
lifespan) would requi   on of the human population would have to
devote its work almos

One way around the   e oxygen, food, water, and other support
systems we require—   bitable planet where, upon arrival, they
would be reconstruc   human DNA, the robots would essentially
"grow" a human co   is one idea under consideration by the
100 Year Starship (
project of the NAS/
Center and the De
Research Projects
which was establi
U.S. Department

DARPA is known
the-box thinking
the folks largely
the Internet from
practical mecha
between militar
project is DARF
achieve practic
within 100 yea
point out that,
fiction author
conceive of th
send a man t   *years after Jules Verne wrote* From the Earth to the Moon,
over a hundr   *me reality when Neil Armstrong and Buzz Aldrin (pictured)*
States made   *people to set foot on the Moon.*

The Hubble Space Telescope captured this image of the Egg nebula, a pre-planetary nebula. At its center, thick dust obscures the star that is rapidly casting off shells of gas and dust as it transforms itself into a white dwarf.

THE WHITE DWARF PHASE
(11 BILLION TO ABOUT 10 QUADRILLION YEARS)

The Sun's red giant phase will essentially end when gravity compresses the Sun's core so much that the atomic structure of the core's mass begins to break down. What remains will be crushed into what is known as degenerate matter, a stew of subatomic particles (electrons, neutrons, protons, and so on) incapable of interacting with one another in the same fashion as normal matter. This collapse will release enormous amounts of energy, heating the core to a temperature of about 1 billion degrees Celsius (over 1.8 billion degrees Fahrenheit).

This critical temperature will ignite the fusion of helium into carbon. But instead of the slow, steady burn of hydrogen fusion, the degenerate state of matter in the collapsed core will cause the entire core to ignite nearly simultaneously in what is known as a helium flash. Then a process similar to the red giant phase (but much quicker) will begin. Helium in the central core will be fused into carbon at a slightly faster rate than helium at the outer edges, and the fusion process will move outward to form a burning shell of helium surrounding a carbon-oxygen core. What remains of the ionized plasma making up the Sun's corona will be thrown off as a glowing halo of solar matter known as a stellar remnant nebula (sometimes called a planetary nebula). This is known as the asymptotic giant phase.

Our Sun does not have enough mass to produce gravitational forces strong enough to fuse carbon in a runaway reaction that would spark a supernova. Instead, after shedding its outer layers, all that will remain is a simmering core of carbon and oxygen known as a white dwarf. The Sun will no longer undergo fusion and so will no longer have a source of energy. Its core will have collapsed as much as it can. Although the Sun will still be very hot even as a white dwarf, from this point on, its remaining energy will radiate away as it gradually cools down.

Because its interior will be mostly degenerate matter, the Sun will have very little atomic activity and will maintain a constant temperature of about 100 million degrees Celsius (about 1.8 million degrees Fahrenheit). Its outer shell, however, will cool to about 100,000 degrees Celsius (a little over 180,000 degrees Fahrenheit). But because the Sun's remaining heat will radiate from such a smaller surface area and the cooling process will gradually slow, this phase could last a very, very long time. There are even some white dwarfs almost as old as the known universe still radiating at temperatures of a few thousand degrees.

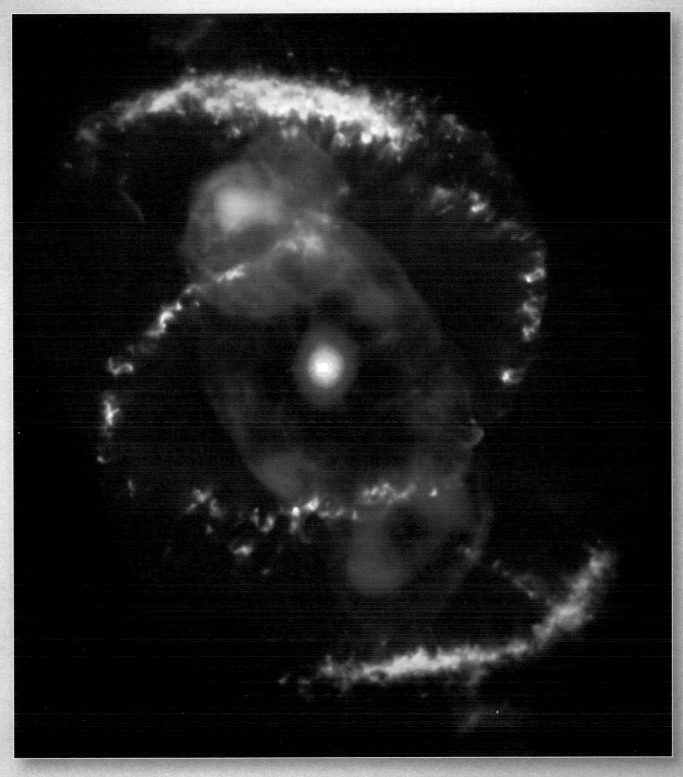

This composite image from the Chandra X-ray Observatory and the Hubble Space Telescope shows the white dwarf at the center of the Cat's Eye planetary nebula.

"LIFE, FOREVER DYING TO BE BORN AFRESH,
FOREVER YOUNG AND EAGER, WILL PRESENTLY
STAND UPON THIS EARTH, AS UPON A FOOTSTOOL,
AND STRETCH OUT ITS REALM AMIDST THE STARS."

—*H.G. WELLS*, THE OUTLINE OF HISTORY, *1920*

THE BLACK DWARF PHASE
(>10 QUADRILLION YEARS, IN THEORY)

Scientists believe the known universe is about 13.79 billion years old. Since nothing can be older, we can only predict what our Sun will look like after 13.79 billion years. Theoretically, it will continue to cool until its temperature is equal to its surroundings. At this phase, when the Sun stops radiating heat altogether, it will have become a black dwarf.

No black dwarfs are thought to exist yet. Scientists speculate they might look very similar to a planet but exert substantially greater gravitational pull on nearby objects. In fact, if we ever find a black dwarf, it will probably be because we have detected distortions caused by its gravitational pull on objects around it.

In this image of the ancient globular star cluster NGC 6397 from the Hubble Space Telescope, white dwarf stars ranging from less than 800 million years old to about 3.5 billion years old are dotted among the smaller, less brightly shining stars.

LOOKING AHEAD

Despite the doom that awaits humanity as our Sun dims, the future looks bright. We are only just beginning to understand the complex forces within our Sun and how they impact the space environment. Satellites launched just a few years ago are revolutionizing our understanding of the Sun and its complex interconnection with all parts of the solar system. Over the next several years, NASA will launch even more satellites as part of the fleet of spacecraft comprising the Heliophysics Great Observatory.

MAGNETOSPHERIC MULTISCALE MISSION

In August 2014, NASA will launch four identical spacecraft designed to study the Earth's magnetosphere as part of the Magnetospheric Multiscale Mission (MMS). The spacecraft will take critical measurements of the magnetosphere's electron diffusion region, the place where magnetic field lines rearrange and convert magnetic energy to kinetic energy during magnetic reconnection. It is also the mechanism by which energy is transferred from the Earth's magnetic field to charged particles in its upper atmosphere. A better understanding of this process may provide insights into how it works in the Sun's corona. Knowing how magnetic reconnection accelerates charged particles around the Sun may help develop more accurate forecasting of coronal mass ejections and better defenses against their impacts on the Earth.

An artist's concept of the four spacecraft to be launched as part of the Magnetospheric Multiscale Mission

SOLAR ORBITER

By January 2017, NASA plans to launch the Solar Orbiter (SolO), a joint project with the European Space Agency (ESA), designed to make very close observations of the Sun. Carrying 10 precision instruments, SolO will, among other things, analyze the magnetic properties and composition of the solar wind and map magnetic variability within the heliosphere. Scientists have observed, for example, that the density of charged particles within the heliosphere appears to be increasing, but they don't know why. By collecting precise data on the properties of the Sun's magnetic forces, SolO may reveal the processes driving connections between the Sun, the Earth, and the outer reaches of the solar system.

SOLAR PROBE PLUS

In 2018, NASA plans to launch the Solar Probe Plus, a robotic spacecraft designed to probe the Sun's outer corona, passing nearly four times closer to the Sun than any previous spacecraft. Getting that close will accelerate Solar Probe Plus to speeds up to 200 kilometers (about 124 miles) per second, giving it the distinction of being the fastest human-made object ever created. When it reaches its final orbit around the Sun, the probe will be exposed to temperatures higher than 1,400 degrees Celsius (2,600 degrees Fahrenheit). A shield at the front of the probe, made of reinforced carbon-carbon composite, will protect the precision instruments on board.

If all goes as planned, the probe will analyze the forces driving the solar wind, explore how plasma "dust" in the Sun's corona influences the formation of charged particles, and trace the energy flows that accelerate these particles as they follow the solar wind outward. This information may prove invaluable in developing methods of protecting critical GPS and communications satellites orbiting the Earth and vulnerable electronics on its surface.

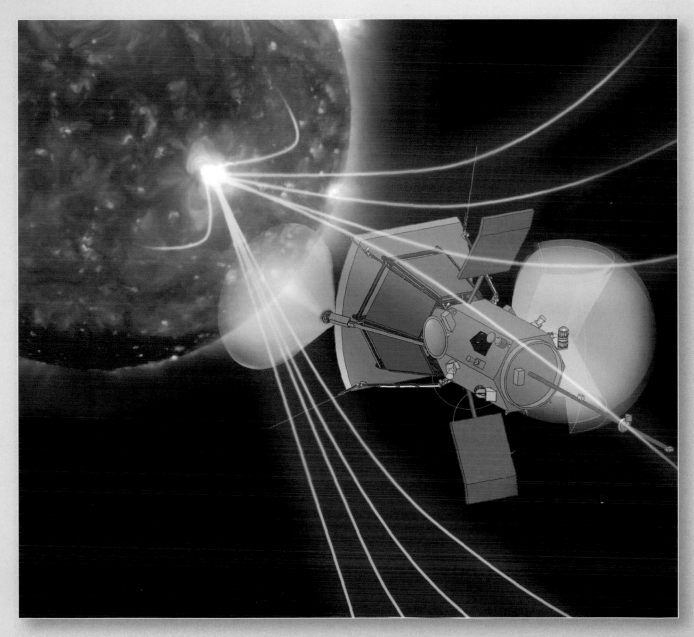

The Solar Probe Plus approaches the outer layer of the Sun in this artist's rendering.

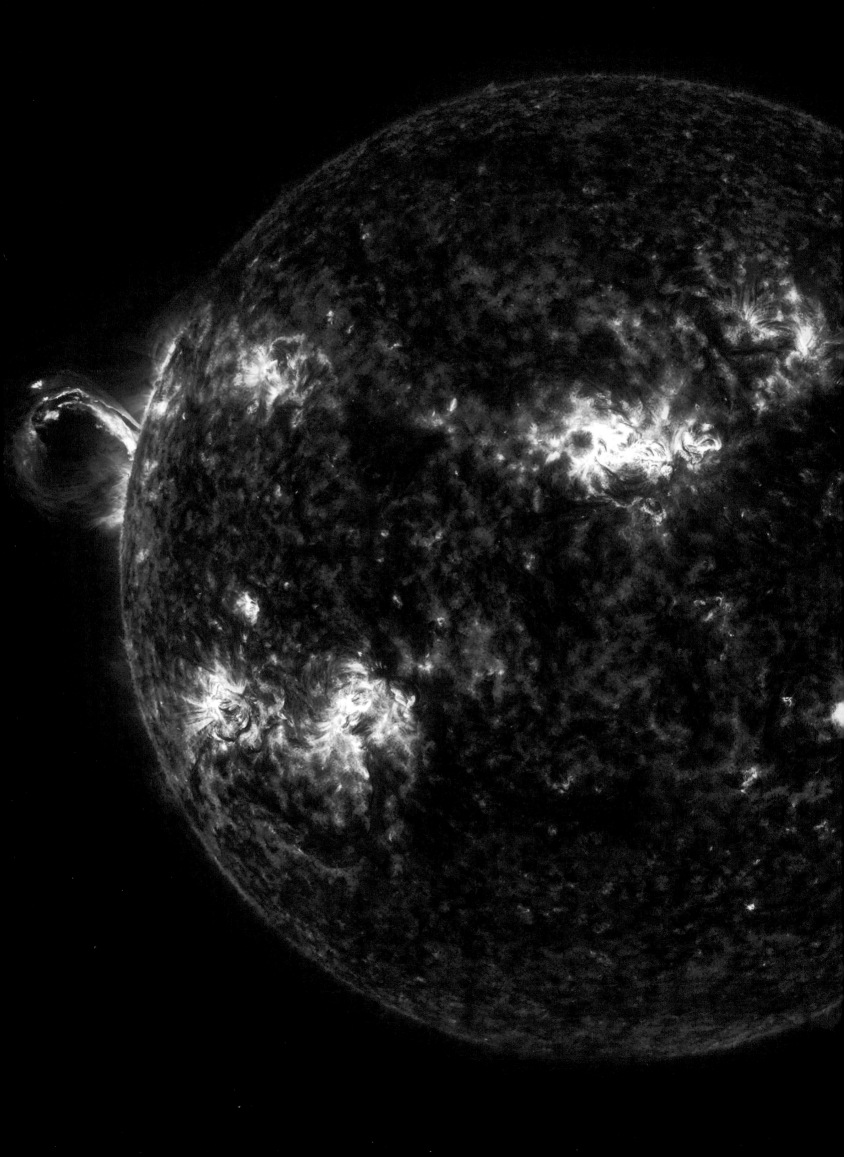

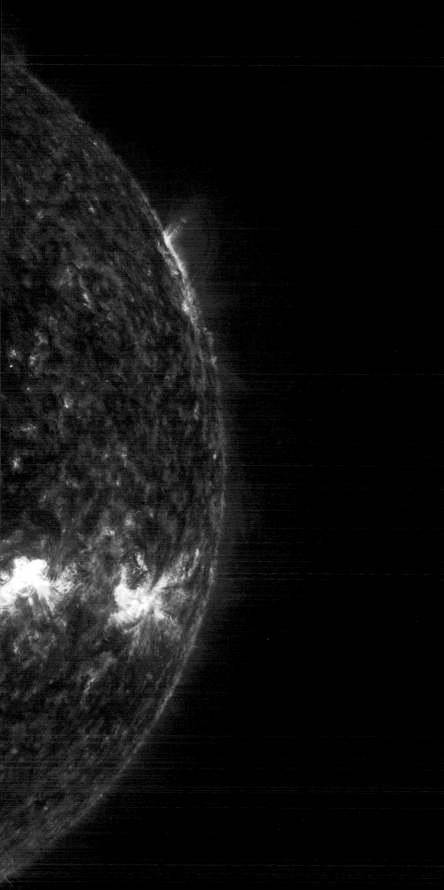

Ironically, the more we learn about the science of our Sun, the more we can appreciate it as something more than the subject of scientific study. When we know how our bodies take energy from the Sun and transform it into vital nutrients, when we comprehend how magnetic fields generated deep within the Sun impact the electronic gadgets in our pockets, when we understand that everything in the solar system exists within its outermost layers, our Sun becomes more than a fixture of "outer space."

The Sun is humanity's constant companion, playing as close a relationship with life on planet Earth as any celestial body could. In its rising and setting, it determines the rhythm of our lives. It bathes us in warmth and gives us the ability to see everything around us—all the goodness and all the tragedy of life. For all the breakthroughs in astronomy and physics the study of our Sun has inspired, perhaps its greatest gift to humanity is in helping us to better understand ourselves and our unique place in the universe.

Our Sun is one among billions of stars. But it is the only star so immediately experienced by all humankind. The Sun connects every person of every race, nationality, and religion. No matter where (or *when*) they have lived, every human being shares the Sun (whether we want to or not). That is something that can be said of very few things. And when we have finally understood everything that makes our Sun tick, when science has solved every mystery, it will still be our shared relationship with the Sun that makes it truly remarkable. Indeed, the Sun is not just a star. It is *our* star . . . and the only star that matters.

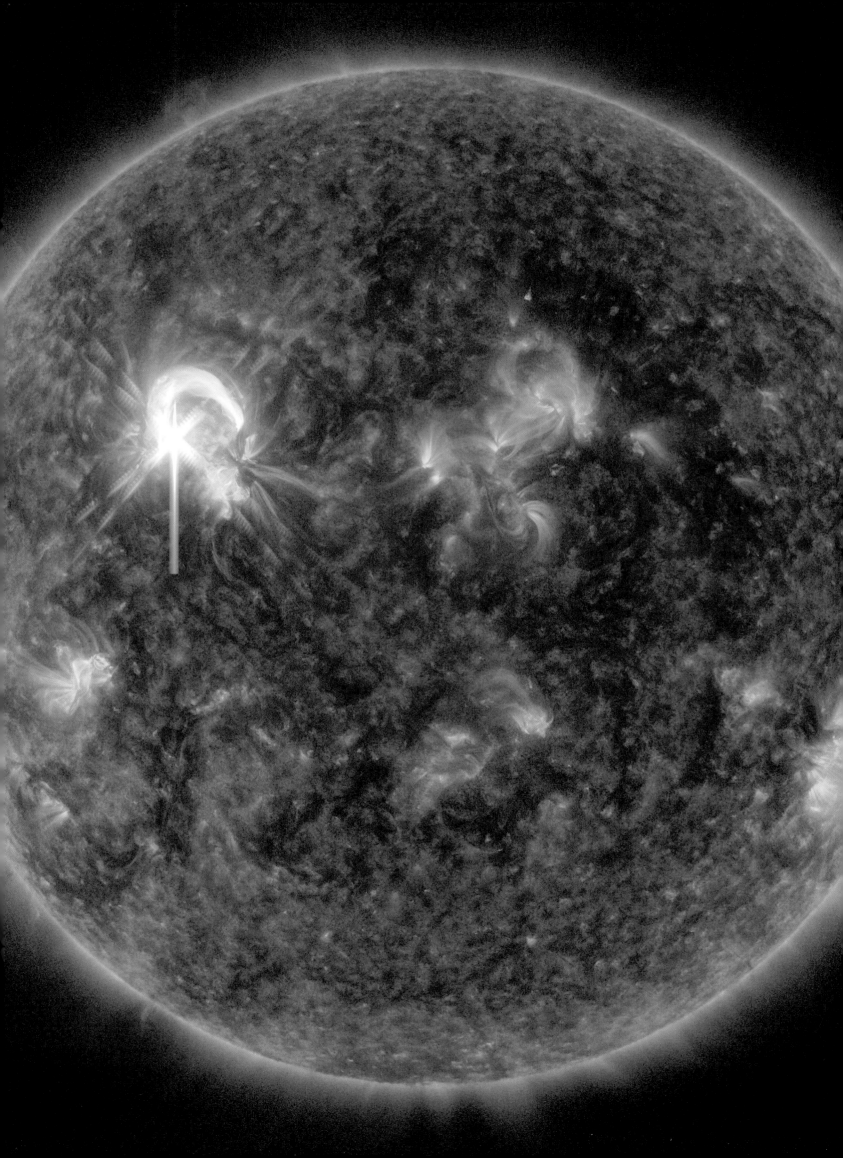

ACKNOWLEDGMENTS

They say the young are fearless because they don't know enough to be afraid. I can't say that I approached putting together a trade book on the science of the Sun fearlessly. But I can say I was a babe in the woods when I started the process. The idea was simple enough, even obvious: NASA's Solar Dynamics Observatory (SDO) is producing photographs of the Sun in such unprecedented detail that someone really ought to publish them in a coffee-table book. What's more, the book should explain the science of the Sun in a way that any casual reader can understand. As an energy policy expert and avowed science geek, I had written several books explaining the technical aspects of electricity and the intricacies of energy markets for amateur readers and imagined that a book simplifying the science of the Sun would be no different.

I had no idea how Herculean a task I was proposing so flippantly. And I'm glad. Had I known the tremendous amount of research, writing, rewriting, editing, fact-checking, rechecking, updating, and coordinating involved, I may never have jumped into the task with such abandon. The Sun is a topic as big as its subject, and information about the Sun is constantly changing (thanks in no small part to new discoveries made possible by NASA's incredible instruments and the extraordinary scientists that use them). I have no doubt that, even with the help of far more talented individuals, I've gotten some of it wrong and, for that, I beg your forgiveness. All mistakes herein are entirely mine.

Any credit, however, goes to an amazing team of people, whose support and assistance turned my blissfully naïve thought into this remarkable book. I am immensely grateful for the support and friendship of Jeff McLaughlin at Race Point Publishing, who shepherded this project from its genesis. His unwavering confidence in me often proved greater than my own. His ability to float effortlessly above the torrent of book development provided much-needed calm in the midst of the storm and is a testament to his unparalleled talent as a publisher.

Many thanks go to Nancy Hall and Linda Falken of The Book Shop, Ltd. As the book's editor, Linda not only placed every dot and tittle to ensure the book's stylistic clarity, she culled through reams of research, meticulously checking for technical accuracy. If not for her dedication to going above and beyond her editorial duties, the text would contain several embarrassing factual errors. If they gave an award for "Most Patience Shown a Trade-Book Newbie," Nancy Hall would easily take the trophy. She coordinated all the simultaneous moving parts involved in a project this complex and allowed the rest of us to focus on meeting very challenging deadlines. In a close second would be graphic artist Tim Palin, whose visually compelling design transformed my dense text into stunning art.

This book began as an idea sparked by NASA's photographs and ended as a kind of homage to this consistently undervalued agency. One way or another, I owe NASA for every word and image in this book. But I am especially grateful to Dr. Steele Hill, media specialist at NASA's Goddard Space Flight Center, who provided invaluable advice on the initial outlines, suggested some of the book's most dazzling images, and facilitated its review by the renowned astrophysicist Dr. Lika Guhathakurta, lead scientist for NASA's "Living With a Star" program. Special thanks also goes to Dr. David Spergel, Princeton's superstar astrophysicist, who contributed the foreword (and whose work on the early universe I have long admired).

Anyone who has written a book knows that no amount of technical and professional help can make up for the unconditional support and daily sacrifices of individuals whose only stake in the project is their love for the author. For me, this generous and long-suffering group includes my parents, Bob and Sharon Cooper, and my touchstone, David Styers. But, most of all it includes my best friend and roommate, Thomas Makely, who endured more than any roommate should ever have to seeing this project to fruition.

—Christopher Cooper

GLOSSARY

A

adenosine triphosphate (ATP) – a product of photosynthesis, ATP is a coenzyme essential for transporting chemical energy within cells.

aerobic respiration – a process of generating cellular energy that uses oxygen as the final electron receptor.

anaerobic respiration – any process of generating cellular energy that uses something other than oxygen as the final electron receptor.

angular momentum – in simple terms, the product of a spinning object's rate of rotation and the amount of matter (mass) that is rotated.

antimatter – matter composed of antiparticles, which have the same mass as subatomic particles in ordinary matter but opposite electrical and magnetic charges. When matter and antimatter come together, they destroy each other.

anisotropy – a measurable difference in property. In astronomy, anisotropy generally refers to the tiny differences in temperature between areas of mass in the infant universe as measured by cosmic background radiation.

Archaea – a domain of single-celled microorganisms, many of which utilize anaerobic respiration, whose cells do not contain nuclei or other membrane-bound intracellular structures.

astrometric – referring to the measurement of the position and motion of celestial bodies; also, a type of binary star system that can be inferred from the motion caused by the gravitational interaction of celestial bodies.

Astronomical Unit (AU) – a unit of length and measurement defined as exactly 149,597,870,700 meters (92,955,807.3 miles), or roughly the mean distance between the Earth and the Sun.

asymptotic giant phase – the stage in the life cycle of a main sequence star after it has fused the helium in its core to carbon and the process of helium fusion moves outward, greatly increasing luminosity.

aurora – a light display in the Earth's atmosphere (usually at higher latitudes) that occurs when charged particles from the solar wind collide with the Earth's magnetosphere, causing them to glow. The aurora observed in the northern hemisphere is called the aurora borealis (the northern lights), while the aurora observed in the southern hemisphere is called aurora australis (the southern lights).

B

baryogenesis – a hypothetical physical process in the early moments of the universe that produced an asymmetry between matter and antimatter, resulting in a residual amount of matter from which everything in the universe formed.

beta-plus decay – a type of radioactive decay in which an atom emits a positron and an electron neutrino from its nucleus, reconfiguring the quarks in a proton and turning it into a neutron.

binary star system – two stars that are so close together that their gravitational interaction causes them to orbit around a common center of mass.

bioluminescence – the production and emission of light by living organisms; also, the light produced.

black dwarf – a hypothetical stage in the life cycle of a star when it cools to the point that it no longer emits significant heat or light. No black dwarfs exist (yet) because the time calculated for a white dwarf star to reach this stage is longer than the current age of the universe.

blue shift – a shift in the wavelengths of an approaching light source toward the ultraviolet end of the electromagnetic spectrum caused by the compression of light waves.

C

calcifediol – a form of stored vitamin D that has the potential to be converted into active vitamin D in our bodies.

Cambrian Explosion – the relatively rapid appearance of most animal phyla around 540 million years ago, accompanied by a massive diversification in the types organisms.

camera obscura – an optical device that projects an image of an object onto an inside surface of a darkened room or box by passing light through a small hole in one side. Camera obscura were often used to observe the Sun without damaging the eyes.

center of mass – the balance point between the gravitational forces of two or more objects where the net gravitation force on either object is zero.

chromosphere – the lower layer of the Sun's atmosphere, located between the photosphere (the Sun's visible surface) and the corona.

circumstellar habitable zone – the region around a star within which it is theoretically possible for a planet to maintain liquid water on its surface and, therefore, be capable of supporting life.

concave lens – a lens that is thinner at the center and thicker around the edges so that that light passing through it converges, causing objects to appear smaller.

conservation of angular momentum – a law of physics that holds that the angular momentum generated by spinning an object cannot be created or destroyed, only transferred. The law explains why fluid spinning objects like the Sun exhibit differential rotation.

convection – the transfer of thermal energy by the movement of fluids, usually through convection currents.

convection cell – where plasma circulates in the Sun's convective zone, with hot plasma rising to the surface, cooling, and sinking back inward.

convection current – the movement of plasma from the Sun's hotter interior to its cooler surface and then back again.

convective zone – the layer of the Sun between the radiative zone and the photosphere in which energy is transferred primarily through convection.

convex lens – a lens that is thicker at the center and thinner around the edges so that that light passing through it spreads out, causing objects to appear larger.

Copernican Revolution – a paradigm shift in scientific thinking from a geocentric model of the universe (where the Earth is at the center) to a heliocentric model (where the Sun is at the center), generally attributed to the work of astronomer Nicolaus Copernicus.

Coriolis effect – the apparent deflection toward the direction of rotating objects moving through a rotating frame, so that, from a static position, it appears some force has been applied to the object, bending its path.

corona – the extremely hot upper layer of the Sun's atmosphere, which is composed of low-density plasma that extends for millions of kilometers into space.

coronal mass ejection (CME) – the ejection of charged material from the Sun's atmosphere, usually caused by a sudden release of energy when loops of magnetic field lines near the Sun's surface snap open.

cosmic inflation – the theoretical rapid expansion of the universe at rates exceeding the speed of light, during the first fractions of a second after the Big Bang.

cyanobacteria – a phylum of bacteria, commonly known as blue-green algae, that obtain their energy through photosynthesis.

D

dalton – the standard unit for indicating mass at an atomic scale, named after the nineteenth-century English chemist John Dalton, a pioneer of modern atomic theory.

deferent – in the Ptolemaic model, a large ring around the Earth, along which a planet traveled as it rotated on its epicycle.

degenerate matter – a collection of free-floating, non-interacting atomic particles (electrons, neutrons, protons, etc.) created by the extraordinarily high density within compacted stars and star remnants.

differential rotation – a phenomenon in fluid spinning objects (like the Sun) in which angular momentum within the object is not equally distributed and different parts rotate at different rates.

diproton – an unstable isotope of helium, also known as helium-2, that has two protons and no neutrons, and usually decays quickly into two separate protons.

doping – in solar photovoltaics, the process of intentionally introducing impurities into silicon to manipulate its conductivity and induce the flow of electrons.

Doppler effect – the apparent change in frequency by the contraction or expansion of waves (including sound and light) as the source emitting them moves closer to or farther from the person detecting them.

dwarf star – any main sequence star with luminosity within a certain range; yellow dwarfs are stars of this type that have a mass comparable to the Sun's.

dynamic pressure – the energy per unit volume of a fluid as it is compressed, as in the pressure exerted on the leading edge of an airplane wing as it moves through the atmosphere.

E

electromagnetic spectrum – the range of all possible frequencies (or wavelengths) of electromagnetic radiation, from radio waves (with lengths from hundreds of meters to about 1 millimeter) to short gamma rays (with wavelengths below 1-trillionth of a meter).

electromagnetism – one of the four known fundamental forces of nature, it derives from the interaction of electrons between molecules and can induce an electrical charge and/or a magnetic field.

electron transport chain – in photosynthesis, the transfer of electrons between donor molecules and acceptor molecules across a cellular or intracellular membrane.

emission spectrum – the range in frequencies of electromagnetic radiation emitted by an element or chemical compound after its atoms have been excited, such as by burning.

epicycle – in the Ptolemaic model, a circle on which a planet rotated. The epicycle orbited the Earth on a larger ring called a deferent. Epicycles were devised by Ptolemy to account for the apparent retrograde motion of some planets.

Era of Recombination – a period in the early formation of the universe during which matter cooled enough that electrons became bound to protons, forming neutrally charged hydrogen atoms and allowing photons to travel without being scattered by free-floating particles.

exoplanet – any planet outside our solar system, the presence of which is usually inferred by variations in a star's brightness that occur when the exoplanet's orbit brings it between the star and the Earth.

F

focal point – the point at which rays of light converge after bouncing off of a concave mirror.

Fraunhofer lines – a set of dark vertical lines in the optical spectrum of a light source that corresponds to various elements comprising the source.

fusion – a nuclear reaction during which the nuclei of atoms under immense heat and pressure join to form a new nucleus, converting a small amount of mass to energy in the process.

G

gamma-ray burst – a narrow beam of intense radiation, seen as bright flashes, released by hypernovas during the collapse of massive stars.

geocentrism – a model of the cosmos in which the Earth is at the center and all celestial bodies orbit around it.

Gleissberg cycle – a proposed fluctuation in the strength of solar cycles over a period of about 87 years, named after the astronomer Wolfgang Gleissberg.

granule – the top of a convection cell where heated solar plasma reaches the Sun's photosphere.

H

heliocentrism – a model of the solar system in which the Earth and planets revolve around the Sun.

heliomagnetic reversal – the 11-year cyclical change in the Sun's magnetic field during which its poles switch polarity.

heliophysics – study of the matter and motion of the Sun and its magnetic field.

helioseismology – the study of wave oscillations in the Sun to provide information about matter and motion within it.

hydrogen burning – an expression often used to refer to the process of nuclear fusion during which hydrogen is fused to form helium.

hydrogen depletion – the point at which a star stops fusing hydrogen into helium.

hypernova – an intense supernova explosion thought to create gamma-ray bursts as a massive, fast-spinning star collapses to form a black hole.

I

induced current – an electrical charge through terrestrial conductors (usually wires, transformers, and circuitry) created when electromagnetic energy from the Sun interacts with the Earth's geomagnetic field.

International Unit (IU) – a standardized unit of measurement of a substance based on its biological activity.

interplanetary magnetic field – the solar magnetic field carried by the solar wind outward through the solar system.

ionization – the addition or removal of an ion (a charged particle such as an electron) from an atom.

isotope – any of a type of element that varies by the number of neutrons in its atomic nuclei.

L

lodestone – magnetite, a naturally magnetized mineral.

M

macrophage – a specialized cell of the immune system that engulfs and digests cellular debris and pathogens.

magnetic flux – a measure of the amount of a magnetic field that passes through a surface.

magnetic reconnection – a process in highly conductive plasmas in which magnetic field lines from different magnetic domains are spliced together or otherwise change their patterns of connectivity with respect to the source of the field.

magnetosphere – the protective outer layer of the Earth's ionosphere that deflects electromagnetism and charged particles from the Sun.

main sequence star – a distinctive type of star, also called a dwarf star, with color and brightness similar to our Sun.

Maragha Revolution – the paradigm shift away from the Ptolemaic cosmological model by thirteenth-century Persian astronomers associated with the observatory near Maragha (also known as Maragheh), Iran.

Maunder Minimum – a prolonged period of reduced sunspot activity, from about 1645 to 1715, during which Europe experienced lower-than-average temperatures.

megaelectron volt – a measure of energy equal to one million times the amount of energy gained by an electron as it accelerates through one volt.

N

nebula – an interstellar cloud of dust, hydrogen, helium, and ionized gas from which stars form.

neutrino – a tiny, electrically neutral subatomic particle released during the process of hydrogen fusion inside the Sun.

nucleosynthesis – a process that began about three minutes after the start of the Big Bang in which the first atomic nuclei were formed by the combination of protons and neutrons.

O

Oort Cloud – a hypothesized cloud made of billions of icy dust clumps surrounding the solar system at a distance of about one light year from the center.

opsin – any of the light-sensitive proteins found in photoreceptor cells that help convert photons into an electrochemical signal.

osteomalacia – a painful softening of the bones (commonly known as rickets) often resulting from a vitamin D deficiency.

Oxygen Catastrophe – the abrupt increase in the concentration of atmospheric oxygen about 200 million years after the first cyanobacteria started expelling it as a by-product of photosynthesis; also known as the Great Oxygen Event.

oxygenic photosynthesis – the conversion of sunlight into chemical energy, which produces oxygen as a by-produce and releases it into the atmosphere.

P

parallax – the difference in the apparent position of an object viewed along two different lines of sight. Astronomers use parallax to calculate the distance of celestial objects.

photochemical – a chemical reaction initiated by the absorption of energy in the form of light.

photon – a packet of light or other electromagnetic radiation

photosphere – the deepest region of a star from which light is radiated, it is the turbulent "surface" of our Sun.

photovoltaic – relating to a method of converting sunlight into electricity by exciting the electrons in a conductive material, which causes them to jump free of their atoms.

plasma – one of the four fundamental states of matter, it is composed of charged particles (ions) that are highly reactive to electromagnetism.

polarity – a property of an electromagnetic object that helps describe the charge to which it is attracted. Conventionally magnetic field lines emanate from the "north" pole and re-enter through the "south" pole.

positron – the antimatter counterpart of an electron, often generated by the radioactive decay of a substance; also known as an antielectron.

Precambrian Era – the time period from the formation of the Earth (about 4.54 billion years ago) until the beginning of the Cambrian period (about 541 million years ago), which was dominated by life forms utilizing anaerobic respiration to create cellular energy.

precession, precessional movement – the wobble of the Earth on its axis over a 26,000-year rotation that is thought to cause the position of the stars to appear to shift slowly over time.

prominence – a filament of charged plasma (often looped) anchored at the Sun's photosphere and extending into the chromosphere and corona.

proton-proton chain reaction – a type of fusion reaction that converts hydrogen into helium by first forming deuterium (a hydrogen atom with one proton and one neutron in its nucleus) and then fusing deuterium atoms.

protostar – the earliest stage in the formation of a star when an area of mass grows inside a cloud of interstellar dust. Our Sun was a protostar for the first 100,000 years of its life.

Ptolemaic model – a geocentric model of the universe, proposed by the astronomer Claudius Ptolemaeus, that utilizes deferents and epicycles to explain the apparent retrograde motion of some planets.

Q

quantum mechanics – a branch of physics that attempts to describe phenomena at the subatomic level, incorporating various concepts including the wave-particle duality of light and matter.

quark – a fundamental constituent of matter. The configuration of quarks determines the electrical charge of their composite particles (such as protons and neutrons).

R

radiation – the transfer of energy through electromagnetic waves that travel between atoms.

radiative zone – a layer in the Sun's interior between the core and the convective zone in which energy is transferred primarily through radiation.

radioisotope – a type of isotope with an unstable nucleus that emits radiation at a steady rate as it decays to a stable form.

red giant – a late phase in the evolution of some stars, when the star has converted all the hydrogen in its core to helium, and hydrogen fusion speeds outward in a shell surrounding the core.

red shift – a shift in the wavelengths of a receding light source toward the infrared end of the electromagnetic spectrum caused by the expansion of light waves due to the Doppler effect.

retrograde motion – the apparent reversal in motion of a planet or other celestial object as observed from Earth.

S

shearing – a force derived from the difference in the movement of particles within a continuous material. In the Sun, it is thought to arise in the tachocline because of the difference in angular velocity of plasma in the radiative zone relative to plasma in the convective zone.

sidereal rotation – the apparent 24.47-day rotation period of the Sun as observed from the Earth. It is shorter than the Sun's actual (synodic) rotation period due to the Earth's rotation around the Sun.

solar cycle – the 22-year cyclical change in solar activity as measured by the number of sunspots and degree of solar radiance. The cycle consists of alternating 11-year periods of solar maximum and solar minimum.

solar flare – the sudden release of electromagnetic radiation from the Sun due to the rapid superheating of plasma in the Sun's atmosphere. Flares are usually accompanied by a bright flash at the Sun's surface.

solar maximum – an 11-year period of the solar cycle characterized by an increase in solar activity as measured by the number of sunspots and the degree of solar radiance.

solar minimum – an 11-year period of the solar cycle characterized by a marked decrease in solar activity as measured by the number of sunspots and the degree of solar radiance.

solar wind – a continuous emission of charged particles from the upper layers of the Sun's atmosphere that interacts with the Earth's magnetosphere to produce beautiful auroras.

spectral signature – the specific combination of electromagnetic radiation of various wavelengths emitted by a source of light that, through spectroscopy, can indicate its composition and relative motion.

spectroscopy – the use of spectral signatures to study the interaction between matter and radiated energy. It can be used to determine the composition and relative motion of stars and other celestial bodies.

sunquake – seismic waves in the Sun's photosphere usually caused by solar plasma in prominences falling back into the Sun's surface with extreme force.

sunspot – a dark, relatively cool area on the Sun's photosphere where intense magnetic activity temporarily blocks convection of solar material. Sunspots usually appear in pairs with opposite polarities.

supernova – the violent explosion of a star as gravity collapses the structure of atoms at its center, releasing in a short blast as much energy as our Sun will emit in its lifetime.

synodic rotation – the 26.24-day rotation period of the Sun measured by the time it takes a fixed feature at its equator to rotate to the same apparent position as seen from Earth.

T

tachocline – the relatively thin transition region between the Sun's radiative and convective zones. Only 28,000-kilometers (about 17,000 miles) thick, the tachocline is thought to be the source of much of the Sun's magnetic field.

telescopic projection – a method, first developed by the astronomer Benedetto Castelli, of viewing the Sun by projecting its image through a telescope and onto a flat surface.

terawatt – a measure of power equivalent to one trillion watts. The total power consumed by humans in 2010 was about 16 terawatts.

thermal runaway – a scenario in which the additional thermal energy created by the pressure and gravity of a star's mass will spark carbon fusion, resulting in successive, accelerating types of fusion that increasingly condense the star's core until it collapses upon itself, causing a supernova.

U

ultraviolet (UV) spectrum – a range of electromagnetic radiation with wavelengths shorter than violet visible light but longer than X-rays that can damage human DNA and cause cancer.

V

vector – regarding force, a measure both of its magnitude and of the direction in which it is influencing an object.

W

Waldmeier effect – the proposed inverse relationship between the intensity of a solar cycle and the time it takes for the sunspot number to rise from minimum to maximum; for example, the number of sunspots will increase more rapidly during a strong sun cycle compared to a weak sun cycle.

white dwarf – a late phase in the life cycle of a main sequence star, when all fusion reactions have ceased and the remaining inert carbon-oxygen core emits stored thermal energy while it slowly dims, until the star reaches the age of about a quadrillion years.

Wolf number – a standardized measurement of the number of observed sunspots and groups of sunspots, factoring for the instrumentation used to make the observation (and its location).

INDEX

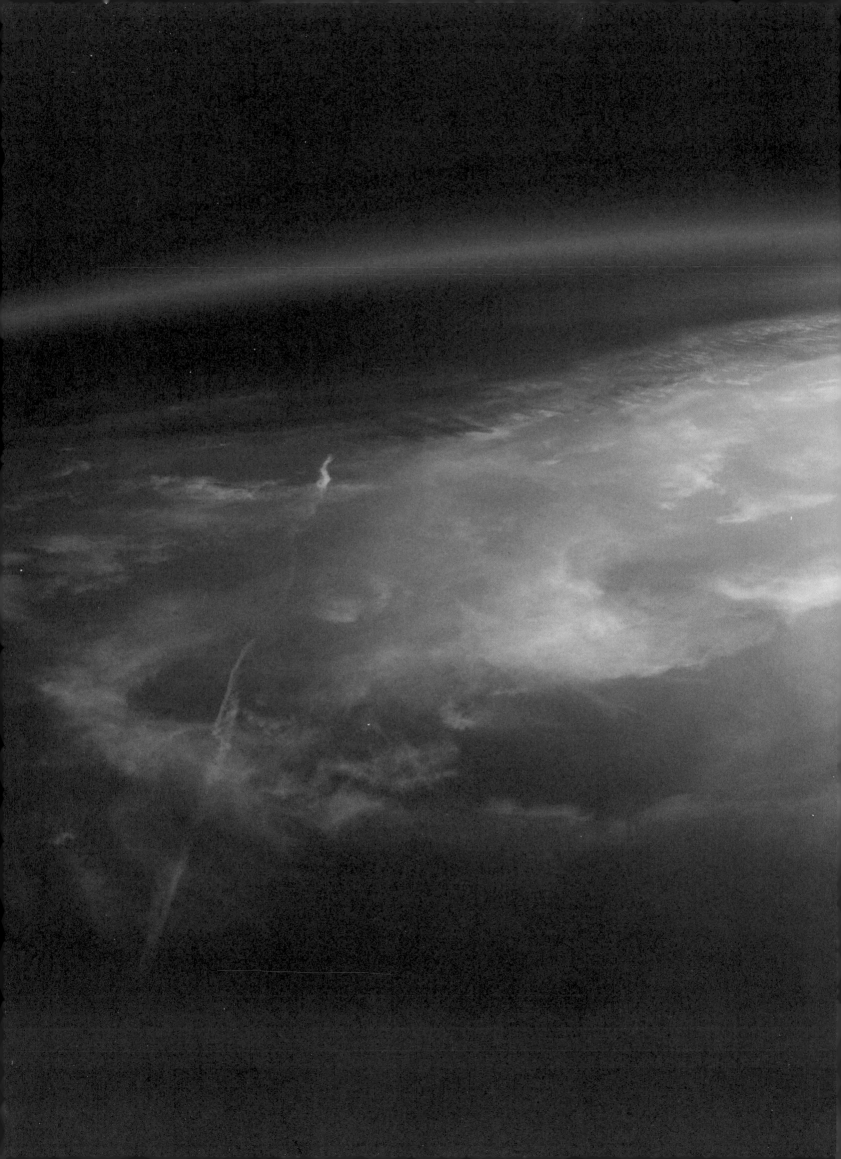